BROKEN
CODE

BROKEN CODE

INSIDE FACEBOOK AND THE FIGHT TO
EXPOSE ITS HARMFUL SECRETS

JEFF HORWITZ

DOUBLEDAY NEW YORK

www.doubleday.com

DOUBLEDAY and the portrayal of an anchor with a dolphin are
registered trademarks of Penguin Random House LLC.

Jacket composite images: senata; miakievy; photopsist; all Getty Images
Jacket design by Pete Garceau

Library of Congress Cataloging-in-Publication Data
Names: Horwitz, Jeff (Journalist), author.
Title: Broken code : inside Facebook and the fight to
expose its harmful secrets / Jeff Horwitz.
Description: First edition. | New York : Doubleday, [2023] |
Includes bibliographical references.
Identifiers: LCCN 2023016222 (print) | LCCN 2023016223 (ebook) |
ISBN 9780385549189 (hardcover) | ISBN 9780385549196 (ebook) |
ISBN 9780385550437 (open-market)
Subjects: LCSH: Facebook (Firm)—History. | Facebook (Firm)—
Corrupt practices. | Social media—United States—History.
Classification: LCC HM743.F33 H67 2023 (print) |
LCC HM743.F33 (ebook) | DDC 302.30285—dc23/eng/20230523
LC record available at https://lccn.loc.gov/2023016222
LC ebook record available at https://lccn.loc.gov/2023016223

MANUFACTURED IN THE UNITED STATES OF AMERICA

1 3 5 7 9 10 8 6 4 2

First Edition

BROKEN
CODE

1

Arturo Bejar's return to Facebook's Menlo Park campus in 2019 felt like coming home. The campus was bigger than when he'd left in 2015—Facebook's staff doubled in size every year and a half—but the atmosphere hadn't changed much. Engineers rode company bikes between buildings, ran laps on a half-mile trail through rooftop gardens, and met in the nooks of cafés that gave Facebook's yawning offices a human scale.

Bejar was back because he suspected something at Facebook had gotten stuck. In his early years away from the company, as bad press rained down upon it and then accumulated like water in a pit, he'd trusted that Facebook was addressing concerns about its products as best it could. But he had begun to notice things that seemed off, details that made it seem like the company didn't care about what its users experienced.

Bejar couldn't believe that was true. Approaching fifty, he considered his six years at Facebook to be the highlight of a tech career that could only be considered charmed. He'd been a Mexico City teenager writing computer games for himself in the mid-1980s when he'd gotten a chance introduction to Apple co-founder Steve Wozniak, who was taking Spanish lessons in Mexico.

After a summer being shown around by a starstruck teenage tour guide, Wozniak left Bejar an Apple computer and a plane ticket to come visit Silicon Valley. The two stayed in touch, and Wozniak paid for Bejar to earn a computer science degree in London.

"Just do something good for people when you can," Wozniak told him.

Success followed. After working on a visionary but doomed cybercommunity in the 1990s, Bejar spent more than a decade as the "Chief Paranoid" in Yahoo's once-legendary security division. Mark Zuckerberg hired him as a Facebook director of engineering in 2009 after an interview held in the CEO's kitchen.

Though Bejar's expertise was in security, he'd embraced the idea that safeguarding Facebook's users meant more than just keeping out criminals. Facebook still had its bad guys, but the engineering work that Facebook required was as much social dynamics as code.

Early in his tenure, Sheryl Sandberg, Facebook's chief operating officer, asked Bejar to get to the bottom of skyrocketing user reports of nudity. His team sampled the reports and saw they were overwhelmingly false. In reality, users were encountering unflattering photos of themselves, posted by friends, and attempting to get them taken down by reporting them as porn. Simply telling users to cut it out didn't help. What did was giving users the option to report *not liking* a photo of themselves, describing how it made them feel, and then prompting them to share that sentiment privately with their friend.

Nudity reports dropped by roughly half, Bejar recalled.

A few such successes led Bejar to create a team called Protect and Care. A testing ground for efforts to head off bad online experiences, promote civil interactions, and help users at risk of suicide, the work felt both groundbreaking and important. The only reason Bejar left the company in 2015 was that he was in the middle of a divorce and wanted to spend more time with his kids.

Though he was away from Facebook by the time the company's post-2016 election scandals started piling up, Bejar's six years there instilled in him a mandate long embedded in the company's official code of conduct: "assume good intent." When friends asked him about fake news, foreign election interference, or purloined data, Bejar stuck up for his former employer. "Leadership made mistakes, but when they were given the information they always did the right thing," he would say.

But, truth be told, Bejar didn't think of Facebook's travails all that much. Having joined the company three years before its IPO, money wasn't a concern, and Bejar was busy with nature photography, a series of collaborations with the composer Philip Glass, and restoring cars with his daughter Joanna, who at fourteen wasn't yet old enough to drive. She documented their progress restoring a Porsche 914—a 1970s model derided for having the aesthetics of a pizza box—on Instagram, which Facebook had bought in 2012.

Joanna's account became moderately successful, and that's when things got a little dark. Most of her followers were enthused about a girl getting into car restoration, but some showed up with rank misogyny, like the guy who told Joanna she was getting attention "just because you have tits."

"Please don't talk about my underage tits," Joanna Bejar shot back before reporting the comment to Instagram. A few days later, Instagram notified her that the platform had reviewed the man's comment. It didn't violate the platform's community standards.

Bejar, who had designed the predecessor to the user-reporting system that had just shrugged off the sexual harassment of his daughter, told her the decision was a fluke. But a few months later, Joanna mentioned to Bejar that a kid from a high school in a neighboring town had sent her a picture of his penis via an Instagram direct message. Most of Joanna's friends had already received similar pics, she told her dad, and they all just tried to ignore them.

Bejar was floored. The teens exposing themselves to girls who they had never met were creeps, but they presumably weren't whipping out their dicks when they passed a girl in a school parking lot or in the aisle of a convenience store. Why had Instagram become a place where it was accepted that these boys occasionally would—or that young women like his daughter would have to shrug it off?

Bejar's old Protect and Care team had been renamed and reshuffled after his departure, but he still knew plenty of people at Facebook. When he began peppering his old colleagues with questions about the experience of young users on Instagram, they responded by offering him a consulting agreement. Maybe he could help with some of the things he was concerned about, Bejar figured, or at the very least answer his own questions.

That was how Arturo Bejar found himself back on Facebook's campus. Eager and highly animated—Bejar's reaction to learning something new and interesting is a gesture meant to evoke his head exploding—he had unusual access due to his easy familiarity with Facebook's most senior executives. Dubbing himself a "free-range Mexican," he began poring over internal research and setting up meetings to discuss how the company's platforms could better support their users.

The mood at the company had certainly darkened in the intervening four years. Yet, Bejar found, everyone at Facebook was just as smart, friendly, and hardworking as they had been before, even if no one any longer thought that social media was pure upside. The company's headquarters—with its free laundry service, cook-to-order meals, on-site gym, recreation and medical facilities—remained one of the world's best working environments. It was, Bejar felt, good to be back.

That nostalgia probably explains why it took him several months to check in on what he considered his most meaningful contribution to Facebook—the revamp of the platform's system for reporting bad user experiences.

It was the same impulse that had led him to avoid setting up meetings with some of his old colleagues from the Protect and Care team. "I think I didn't want to know," he said.

Bejar was at home when he finally pulled up his team's old system. The carefully tested prompts that he and his colleagues had composed—asking users to share their concerns, understand Facebook's rules, and constructively work out disagreements—were gone. Instead, Facebook now demanded that people allege a precise violation of the platform's rules by clicking through a gauntlet of pop-ups. Users determined enough to complete the process arrived at a final screen requiring them to reaffirm their desire to submit a report. If they simply clicked a button saying "done," rendered as the default in bright Facebook blue, the system archived their complaint without submitting it for moderator review.

What Bejar didn't know then was that, six months prior, a team had redesigned Facebook's reporting system with the specific goal of reducing the number of completed user reports so that Facebook wouldn't have to bother with them, freeing up resources that could otherwise be invested in training its artificial intelligence–driven content moderation systems. In a memo about efforts to keep the costs of hate speech moderation under control, a manager acknowledged that Facebook might have overdone its effort to stanch the flow of user reports: "We may have moved the needle too far," he wrote, suggesting that perhaps the company might not want to suppress them so thoroughly.

The company would later say that it was trying to improve the quality of reports, not stifle them. But Bejar didn't have to see that memo to recognize bad faith. The cheery blue button was enough. He put down his phone, stunned. This wasn't how Facebook was supposed to work. How could the platform care about its users if it didn't care enough to listen to what they found upsetting?

There was an arrogance here, an assumption that Facebook's algorithms didn't even need to hear about what users experienced to know what they wanted. And even if regular users couldn't see that like Bejar could, they would end up getting the message. People like his daughter and her friends would report horrible things a few times before realizing that Facebook wasn't interested. Then they would stop.

When Bejar next stepped onto Facebook's campus, he was still surrounded by smart, earnest people. He couldn't imagine any of them choosing to redesign Facebook's reporting features with the goal of tricking users into depositing their complaints in the trash; but clearly they had.

"It took me a few months after that to wrap my head around the right question," Bejar said. "What made Facebook a place where these kinds of efforts naturally get washed away, and people get broken down?"

—

Unbeknownst to Bejar, a lot of Facebook employees had been asking similar questions. As scrutiny of social media ramped up from without and within, Facebook had accumulated an ever-expanding staff devoted to studying and addressing a host of ills coming into focus.

Broadly referred to as integrity work, this effort had expanded far beyond conventional content moderation. Diagnosing and remediating social media's problems required not just engineers and data scientists but intelligence analysts, economists, and anthropologists. This new class of tech workers had found themselves up against not just outside adversaries determined to harness social media for their own ends but senior executives' beliefs that Facebook usage was by and large an absolute good. When ugly things transpired on the company's namesake social network, these leaders pointed a finger at humanity's flaws.

Staffers responsible for addressing Facebook's problems didn't have that luxury. Their jobs required understanding how Facebook could distort its users' behavior—and how it was sometimes "optimized" in ways that would predictably cause harm. Facebook's integrity staffers became the keepers of knowledge that the outside world didn't know existed and that their bosses refused to believe.

As a small army of researchers with PhDs in data science, behavioral economics, and machine learning was probing how their employer was altering human interaction, I was busy grappling with far more basic questions about how Facebook worked. I had recently moved back to the West Coast to cover Facebook for the *Wall Street Journal,* a job that came with the unpleasant necessity of pretending to write with authority about a company I did not understand.

Still, there was a reason I wanted to cover social media. After four years of investigative reporting in Washington, the political accountability work I was doing felt pointless. The news ecosystem was dominated by social media now, and stories didn't get traction unless they appealed to online partisans. There was so much bad information going viral, but the fact-checks I wrote seemed less like a corrective measure than a weak attempt to ride bullshit's coattails.

Covering Facebook was, therefore, a capitulation. The system of information sharing and consensus building of which I was a part was on its last legs, so I might as well get paid to write about what was replacing it.

The surprise was how hard it was to even figure out the basics. Facebook's public explainers of the News Feed algorithm—the code that determined which posts were surfaced before billions of users— relied on phrases like "We're connecting you to who and what matters most." (I'd later learn there was a reason why the company glossed over the details: focus groups had concluded that in-depth explanations of News Feed left users confused and unsettled—the more people thought about outsourcing "who and what matters most" to Facebook, the less comfortable they got.)

In a nod to its immense power and societal influence, the company created a blog called Hard Questions in 2017, declaring in its inaugural post that it took "seriously our responsibility—and accountability—for our impact and influence." But Hard Questions never delved into detail, and after a couple of bruising years of public scrutiny, the effort was quietly abandoned.

By the time I started covering Facebook, the company's reluctance to field reporters' queries had grown, too. Facebook's press shop—a generously staffed team of nearly four hundred—had a reputation for being friendly, professional, and reticent to answer questions. I had plenty of PR contacts, but nobody who wanted to tell me how Facebook's "People You May Know" recommendations worked, which signals sent controversial posts viral, or what the company meant when it said it had imposed extraordinary user-safety measures amid ethnic cleansing in Myanmar. The platform's content recommendations shaped what jokes, news stories, and gossip went viral across the world. How could it be such a black box?

The resulting frustration explains how I became a groupie of anyone who had a passing familiarity with Facebook's mechanics. The former employees who agreed to speak to me said troubling things from the get-go. Facebook's automated enforcement systems were flatly incapable of performing as billed. Efforts to engineer growth had inadvertently rewarded political zealotry. And the company knew far more about the negative effects of social media usage than it let on.

This was wild stuff, far more compelling than the perennial allegations that the platform unfairly censored posts or favored President Trump. But my ex-Facebook sources couldn't offer much in the way of proof. When they'd left the company, they'd left their work behind Facebook's walls.

I did my best to cultivate current employees as sources, sending hundreds of notes that boiled down to two questions: How does a company that holds sway over billions of people actually work? And why, so often, does it seem like it doesn't?

Other reporters did versions of this too, of course. And from time to time we obtained stray documents indicating that Facebook's powers, and problems, were greater than it let on. I had the luck of being there when the trickle of information became a flood.

A few weeks after the 2020 election, Frances Haugen, a mid-level product manager on Facebook's Civic Integrity team, responded to one of my LinkedIn messages. People needed to understand what was going on at Facebook, she said, and she had been taking some notes that she thought might be useful in explaining it.

Haugen was nervous about saying anything further via LinkedIn or on the phone, so we met on a hiking trail in the hills behind Oakland that weekend. After a quarter-mile stroll through California's coastal redwoods, we pulled off the trail to talk in privacy.

Haugen was an unusual source from the start. Facebook's platforms eroded faith in public health, favored authoritarian demagoguery, and treated users as an exploitable resource, she declared at our first meeting. Rather than acknowledging its problems, Facebook was pushing its products into remote, impoverished markets where she believed they were all but guaranteed to do harm.

Since Facebook wasn't dealing with its flaws, she said, she thought she might have to play a role in making them public.

Neither of us had a sufficiently grandiose imagination to guess what that ambition would produce: tens of thousands of pages of confidential documents, showing the depth and breadth of the harm being done to everyone from teenage girls to the victims of

Mexican cartels. The uproar would plunge Facebook into months of crisis, with Congress, European regulators, and average users questioning Facebook's role in a world that seemed to be slipping into ever-growing tumult.

Not every insider I would speak to over the next two years shared Haugen's exact diagnosis of what went wrong at Facebook or her prescription for fixing it. But for the most part they agreed, not just with their fellow corporate turncoats but with the written assessments of scores of employees who *never* spoke publicly. In the internal documents gathered by Haugen, as well as hundreds more provided to me after her departure, staffers documented the demons of Facebook's design and drew up plans to restrain them. Then, when their employer failed to act, they watched one foreseeable crisis unfold after the next.

Whatever the employee handbook might instruct, it was getting harder to assume good intent.

2

"We are going to get blamed for this."

It was late Wednesday morning, the day after the 2016 election, and Facebook's senior Public Policy and Elections staff had gathered in the conference room of their old Washington, DC, office, an unglamorous, cramped space sandwiched among the law and lobbying firms clustered in the city's Penn Quarter. Everyone was trying to understand what Donald Trump's upset victory meant for the company.

Elliot Schrage, Facebook's head of Public Policy and Communications, was the one who offered that grim prediction. Calling in from California, he was convinced Facebook would end up as 2016's scapegoat.

The election had been a rough one for the country—and the company. The rise of Trump brought a new rage to American politics, with racist dog whistles and the crude taunting of opponents becoming a regular feature of mainstream news coverage. The Russians were hacking the Democratic National Committee, while WikiLeaks dumped emails stolen from Hillary Clinton's campaign manager. There was a reason that a meme of a blazing dumpster fire was circulating online when it came to political discussions.

Facebook had already faced its share of criticism. First it was accused by conservatives of censoring trending news stories with a right-wing bent; then came Trump using the platform to launch attacks on Muslim and Mexican immigrants; and, late in the game, the revelation broke that Macedonian hoax sites were fabricating much of the platform's most popular news stories.

The company was due for a reckoning. Like their boss Sheryl Sandberg, long rumored to be under consideration for the job of Treasury Secretary in Hillary Clinton's administration, most of the assembled executives were Democrats. But the prediction that Facebook would take the fall for Trump's election was especially ominous to Katie Harbath, the head of Facebook's Elections team and a Republican. For the past five years, trying to prove that Facebook would transform politics had been her job.

Born to a conservative family in a paper mill town outside of Green Bay, Wisconsin, Harbath had caught the politics bug after volunteering for a Republican Senate campaign in college. After graduating from the University of Wisconsin in 2003, she and a friend moved to DC to look for work. She logged a few months at a Macy's perfume counter before landing an entry-level job at the Republican National Committee.

Harbath arrived at the RNC just as their sole staffer focused on what was then called "e-campaigning" had left. Harbath's only credential was having built a website for a college journalism class, but she put herself forward for what was then considered a marginal post. The response, she recalled, was "'We don't understand this digital stuff—sure, go ahead.'" With the title of associate director of "e-campaigning" and a salary of $25,000, she became a lead player in the Republican Party's online efforts.

Brash and gregarious, Harbath rose quickly. In 2008, she became deputy director for Rudy Giuliani's presidential primary campaign and then went to work for the National Republican Senatorial Committee for the 2010 midterms. Candidates in both parties were looking to emulate the youth-driven, social media–savvy campaign that had carried the nation's first Black president into office, as much because of politicians' desire to associate themselves with social media's cultural cachet as its perceived utility. Merely delivering a speech on Facebook was, at the time, enough to get good press.

Harbath bought a lot of Facebook advertising as part of her job at the NRSC, and she regularly consulted with Adam Conner, who had founded Facebook's DC office in 2007 after working on several Democratic campaigns, including John Kerry's presidential run a few years prior.

By 2011, with another election around the corner, Conner concluded that it wasn't great having Republicans like Harbath discuss advertising strategy with a Democrat like himself—so he asked her to join the company's DC office as one of its first employees.

"I was a little burnt out on the hamster wheel of Republican politics," Harbath said. "By going to a tech company I could do more interesting things than at any campaign." Facebook was everywhere, firmly on its meteoric ascent. *Time* magazine had just named Mark Zuckerberg its 2010 Person of the Year, branding him "The Connector."

Sitting at desks a few feet apart in a Dupont Circle office with an often-broken elevator, Conner and Harbath fielded calls from campaigns seeking digital strategy advice or help after locking themselves out of accounts. But much of the job was simply trying to convince candidates and officeholders that they needed to be on Facebook at all—that the platform wasn't just family photos and fluff. Having the political world pay attention to Facebook was the goal; if the company sold political ads along the way, that was just a bonus.

When the 2012 election was over, Harbath's political team hadn't won—but her corporate one had. At a time when Facebook was looking to compete with Twitter by getting into news and politics, Obama's reelection campaign's prominent use of the platform had been good for Facebook's clout. Zuckerberg wanted to prove that Facebook wasn't just for connecting with friends, and he wanted to do it globally. Harbath, then thirty, became Facebook's global emissary to the political world.

In 2013, she was about to take her first trip—to India, to meet with major political parties. On the eve of her departure, she took out her work journal and titled a page "Indian Phrases." Alongside the words for "hello" and "thank you," she included the phonetic pronunciations of "Lok Sabha" and "Uttar Pradesh." The national legislature and India's largest state, respectively, the words were as basic to Indian politics as "Congress" and "Florida" would be at home.

If Harbath was light on local context or candidate details, it didn't really matter. Facebook was supposed to be a neutral platform, and that meant offering assistance to any major political party that wanted it. Many of India's 1.3 billion citizens were just getting online for the first time via mobile phones, and Facebook wanted to be there when they did. Promoting political ads was a goal, but simply training and encouraging parties to organize on the platform was the main thrust.

Harbath traveled more than half the year, a grueling schedule but one that she found inspiring. She liked the pace, the company, and the mission—Facebook was in a position to bring democracy into the digital age around the world. Social media had already shown its power to topple autocracies during the Arab Spring, and it stood poised to breathe new life into politics worldwide. An internal manifesto from 2012 known as the Red Book declared that "Facebook was not created to be a company" and urged its employees to think more ambitiously than corporate goals. "CHANGING HOW PEOPLE COMMUNICATE WILL ALWAYS CHANGE THE WORLD," the book stated above an illustration of a printing press. Harbath stockpiled multiple copies.

As compelling as Facebook's mission was, the money didn't hurt. Though Harbath joined Facebook less than a year and a half before it went public in 2012, her pre-IPO stock proceeds more than covered the purchase of a two-bedroom condo in Arlington, Virginia. She covered her bathroom in custom wallpaper made out of her Instagram posts. A few featured her drinking beer or on family hunting trips back in Wisconsin. More featured world leaders, photos from work trips, and inspirational posters from Facebook's offices.

If Harbath was drinking the Kool-Aid, so was most of the outside world. The news stories about Facebook's role in politics that she cut out and kept were almost invariably glowing. Politics was such a winner for the company that Facebook's Partnerships team—which worked to boost the presence of publishers, celebrities, and major brands on the platform—tried to subsume it, arguing that politicians were just one more high-profile constituency. Only the intervention of Joel Kaplan, the head of Facebook's Public Policy team in Washington, kept it housed under Harbath.

Not all the excitement was pure hype. Facebook had published research in *Nature* showing it could boost election turnout on a mass scale through messages directing users to state voter registration sign-ups and digital "I Voted" stickers that served as both a reminder to the voter's friends and a subtle form of peer pressure.

Harbath wanted Facebook to do more before the next presidential election—to create dedicated political-organizing tools and channels for elected officials to interact with constituents. Zuckerberg beat her to it. At a 2015 employee town hall, he casually mentioned that Facebook should build an entire team devoted to civic engagement work. To run the new team, initially called Civic Engagement and later Civic Integrity, Facebook poached Samidh Chakrabarti, a digital-democracy advocate who had overseen Google's elections work.

It was on that high note that Harbath began turning her attention to the upcoming U.S. presidential election, a campaign that Facebook hoped would demonstrate that its platform was not just the future of democracy but its present. Armed with statistics showing that the election was already the number one topic on Facebook and case studies on how Facebook could turbocharge fundraising, ad targeting, and voter turnout, Harbath's team sponsored and broadcast every political event it could. When ten candidates showed up to the first raucous Republican presidential primary debate in August 2015, Facebook paid to cosponsor it, broadcasting it live and slapping its logo all over the stage. The company did the same for Democrats, too.

By the spring of 2016, however, Harbath started to feel that something was a little off in online politics. The first sign came not at home but in the Philippines, home to the highest concentration of Facebook users in the world. Ahead of that country's May election, Harbath's team had offered its standard consulting to the major parties. One campaign, that of Rodrigo Duterte, had thrived. A tough-guy mayor, Duterte's presidential campaign was ugly—he cursed out the pope, promised the extrajudicial killing of drug users, and mocked his own daughter as a "drama queen" when she said she had been raped.

Facebook thought Duterte's campaign rhetoric was none of its business—but as the election progressed, the company started receiving reports of mass fake accounts, bald-faced lies on campaign-controlled pages, and coordinated threats of violence against Duterte critics. After years in politics, Harbath wasn't naive about dirty tricks. But when Duterte won, it was impossible to deny that Facebook's platform had rewarded his combative and sometimes underhanded brand of politics. The president-elect banned independent media from his inauguration—but livestreamed the event on Facebook. His promised extrajudicial killings began soon after.

A month after Duterte's May 2016 victory came the United Kingdom's referendum to leave the European Union. The Brexit campaign had been heavy on anti-immigrant sentiment and outright lies. As in the Philippines, the insurgent tactics seemed to thrive on Facebook—supporters of the "Leave" camp had obliterated "Remain" supporters on the platform.

Both votes reinforced Facebook's place in politics—but for Harbath, its role wasn't a feel-good kind. She wasn't going to second-guess voters, but both winning campaigns had relied heavily on Facebook to push vitriol and lies. The tactics' success was all the more uncomfortable given that, back home in the States, a once-long-shot presidential candidate was getting traction with the same playbook.

Donald Trump had made his name in Republican politics by using Twitter to question whether Obama was an American citizen. He surged in the polls after disparaging Mexican immigrants as rapists and mocking Senator John McCain for having been captured and tortured during the Vietnam War. When the primary narrowed to Trump and Texas senator Ted Cruz, Trump went after his opponent's wife and insinuated that Cruz's dad might have helped Lee Harvey Oswald assassinate John F. Kennedy.

Harbath found all that to be gross, but there was no denying that Trump was successfully using Facebook and Twitter to short-circuit traditional campaign coverage, garnering attention in ways no campaign ever had. "I mean, he just has to go and do a short video on Facebook or Instagram and then the media covers it," Harbath had marveled during a talk in Europe that spring. She wasn't wrong: political reporters reported not just the content of Trump's posts but their like counts.

Did Facebook need to consider making some effort to fact-check lies spread on its platform? Harbath broached the subject with Adam Mosseri, then Facebook's head of News Feed.

"How on earth would we determine what's true?" Mosseri responded. Depending on how you looked at it, it was an epistemic or a technological conundrum. Either way, the company chose to punt when it came to lies on its platform.

Facebook had signed on as a sponsor of the Democratic and Republican conventions and threw big parties at both. Harbath handled the Republican convention and was horrified by the speeches from Trump's oddball celebrity acolytes and chants of "Lock her up"—referring to Trump's opponent, Hillary Clinton. When it came time for Facebook's party, none of Harbath's contacts in the Republican establishment were inclined to celebrate. "You have this crazy-ass convention, and it feels like a child's funeral," she recalled.

While Harbath didn't like Trump, Facebook was still in the business of helping both major parties use its products, offering each a dedicated staffer to help target Facebook ads, address technical problems, and liaise with company leadership. The Clinton people turned down the company's offer of an embedded liaison at their Brooklyn headquarters. The Trump people, however, were glad for the help, leaving Harbath with a problem. Nobody on her politics team wanted the gig.

Harbath turned to James Barnes, a Republican friend who worked on the company's political ad sales team. Barnes didn't like Trump any more than Harbath did, but he took on the job in the spirit of a criminal defense lawyer taking on a distasteful client. Resolved to do his best, Barnes relocated to the San Antonio offices of Giles-Parscale, the web marketing firm running Trump's digital campaign. Located by the freeway, across the street from a La-Z-Boy furniture outlet, the office was helmed by the six-foot-eight Brad Parscale, who mainly took on projects building websites for regional clients like the annual San Antonio Rodeo. He had gotten into Trumpworld by building a website for the Trump Hotel brand, then became Trump's digital strategist after building a $1,500 website for his campaign.

Though Trump himself had a knack for social media, Barnes found his campaign needed help, as it was struggling with some of the basics of targeted advertising. Some of the work he did—like figuring out how the campaign could update its targeted Facebook audiences to avoid advertising to absentee voters who had likely cast their ballots already—was novel enough that he and Harbath felt it should be offered to Clinton's camp.

"They said that they didn't have time to integrate that into their processes," Harbath recalled.

Barnes's work in San Antonio ended on October 7, 2016, the day an outtake from *Access Hollywood* leaked. In the footage, Trump, already facing allegations of sexual harassment and assault, boasted about his unsuccessful efforts to sleep with a married woman and declared that his fame meant he could just "grab 'em by the pussy."

Barnes left the office and never came back. Further work with the campaign wasn't just distasteful; it was pointless: the San Antonio campaign office temporarily suspended its ad buys and all but shut down in preparation for defeat. Barnes flew back to Washington, staying in only loose touch with the Trump people. When, just days before the election, he read an article in *Bloomberg Businessweek* containing a boast from Trump's digital team that it was running voter suppression operations on Facebook, he had no idea what they were talking about.

On the evening of the election, Barnes, Harbath, and the rest of the politics team gathered at Facebook's Washington, DC, headquarters. As results rolled in, the group went from celebratory to anxious—to a basement bar near the office. As it became clear Trump was going to win, Barnes took the results especially hard.

"His response was 'Oh my god, what have I done?,' not 'What has the company done?'" Harbath said. "He thought he alone caused this to happen, and he felt incredibly guilty for that."

The next morning, Barnes said, he woke up to a Facebook message from Joel Kaplan's personal assistant. Facebook's top lobbyist wanted a word with Barnes—and, more specifically, to give him a pep talk.

When they met later that day, Kaplan told Barnes that it wasn't his fault that Trump had been elected—or Facebook's fault, for that matter. Kaplan didn't like Trump either, he confided, but he told Barnes he should be proud that he had helped give people a voice. It wasn't for either of them to question what that voice said.

That is what most Facebook executives were telling themselves. Back in Menlo Park, though, much of the company's employee base wasn't having it. Late-afternoon meetings had been canceled as early polling results came in, replaced in some instances by manager check-ins that bordered on grief counseling. By nighttime, when it was clear Trump had won, conversations on Workplace—the secure, internal version of the Facebook site that employees used to work, share documents, and socialize—turned dark and introspective. The core of the company's self-conception was that, by building a platform and connecting people, Facebook was making the world a more knowledgeable and understanding place. Zuckerberg had hammered this idea repeatedly over the years, once suggesting that the adoption of Facebook in the Middle East would end terrorism because connecting people would engender mutual understanding. Although the rise of ISIS, with its online recruiting efforts, put an end to that specific aspiration, the overall talking point survived.

"We've gone from a world of isolated communities to one global community, and we are all better off for it," he had said at the company's developers conference in April 2016, deploring "fearful voices calling for building walls."

Facebook's largely liberal employee base had believed him. Now they weren't really questioning whether Facebook had elected Trump as much as how his victory was compatible with Facebook's existence. Their platform was supposed to have ushered in a new age of social progress and understanding. How had it gone so terribly wrong?

"If we'd had the positive impact we'd intended, this wouldn't have happened," a former senior manager at the company recalled, describing the mood.

Had Facebook turned a blind eye to organized hate efforts on the platform? Did the company have an editorial responsibility to ensure factuality that it hadn't recognized? Were users trapped in "filter bubbles" that prevented them from recognizing Trump as a hateful demagogue?

The same questions were getting asked by journalists, too. Within twenty-four hours of the vote being called, social media was getting walloped in election analysis pieces. Most of the stories weren't uniquely about Facebook—if anything, Twitter loomed larger because of Trump's personal affinity for it. One story, in the *New York Times,* had declared the 2016 election "another Twitter moment"—and dismissed the political import of Facebook and Instagram.

Still, Zuckerberg was angry at the implication that Facebook might have thrown the election. He spoke with Anne Kornblut, the Pulitzer Prize–winning *Washington Post* editor that Facebook had hired the year before to provide it with strategic communications. Facebook, he told her, needed to rebut the criticism as sour grapes.

Zuckerberg believed math was on Facebook's side. Yes, there had been misinformation on the platform—but it certainly wasn't the majority of content. Numerically, falsehoods accounted for just a fraction of all news viewed on Facebook, and news itself was just a fraction of the platform's overall content. That such a fraction of a fraction could have thrown the election was downright illogical, Zuckerberg insisted.

Kornblut, who understood the rhythm of post-election rehashing in the press, urged him to stand down. The last thing Facebook needed was to draw attention to itself at a time when so many people were angry.

But Zuckerberg was the boss. Ignoring Kornblut's advice, he made his case the following day during a live interview at Techonomy, a conference held at the Ritz-Carlton in Half Moon Bay. Calling fake news a "very small" component of the platform, he declared the possibility that it had swung the election "a crazy idea."

"I do think there is a certain profound lack of empathy in asserting that the only reason someone could have voted the way they did is they saw some fake news," Zuckerberg said, adding that the existence of false information on both sides meant that "this surely had no impact."

The merits of the argument weren't immediately material. All that mattered was that, by declaring Facebook to be above reproach, Zuckerberg had painted a target on his company's back. "Mark Zuckerberg denies that fake news on Facebook influenced the elections," read a *Washington Post* article, one of dozens that packaged the CEO's denial with a thorough explanation of the counterargument. ("At least I got him to leave out the numbers," Kornblut later told Harbath.)

Aside from its CEO's do-it-yourself approach to post-election PR, a second problem set Facebook apart from other social media platforms. Trump's campaign really *wanted* to give the company credit for its victory. Nobody but Trump could take credit for his personal Twitter usage, but Parscale had harvested vast sums for the campaign through Facebook's targeted advertising, and he was eager to remind everyone of the coup. "Facebook and Twitter were the reason we won this thing," he told *Wired*.

Gary Coby, the RNC advertising chief who had worked closely with Parscale, went even further, telling *Wired* that the Trump campaign's overall message had been refined by what succeeded in Facebook ads. For good measure, he publicly congratulated Barnes by name on Twitter, calling the Facebook employee the Trump campaign's "MVP." Had Obama's campaign been so effusive about Facebook's power in 2012, Harbath said, champagne corks would have been popping. But the Coby tweet had opened up a can of worms. Not only was his praise of Barnes personally unwelcome; it was going to require Facebook to explain why the hell one of its employees had been embedded with Trump's campaign in the first place.

"We were worried about James's safety," Harbath said. "Meanwhile, Trump's team was coming to us, saying, 'Why aren't you happy we're crediting you?'"

Facebook had at long last gotten the recognition it craved as a central forum for politics—and it was miserable.

A favorite saying at Facebook is that "Data Wins Arguments." But when it came to Zuckerberg's argument that fake news wasn't a major problem on Facebook, the company didn't have any data. As convinced as the CEO was that Facebook was blameless, he had no evidence of how "fake news" came to be, how it spread across the platform, and whether the Trump campaign had made use of it in their Facebook ad campaigns.

Two days after dismissing the possibility that Facebook had inappropriately influenced the election, Zuckerberg retreated. "After the election, many people are asking whether fake news contributed to the result, and what our responsibility is to prevent fake news from spreading," he wrote on Facebook on November 12. "These are very important questions and I care deeply about getting them right." The note stopped short, though, of saying that Zuckerberg thought critics might have a point—he didn't.

Zuckerberg would soon get a taste of how humbling the reckoning was going to be. One week after the election, *BuzzFeed News* reporter Craig Silverman published an analysis showing that, in the final months of the election, fake news had been the most viral election-related content on Facebook. A story falsely claiming that the pope had endorsed Trump had gotten more than 900,000 likes, reshares, and comments—more engagement than even the most widely shared stories from CNN, the *New York Times,* or the *Washington Post.* The most popular falsehoods, the story showed, had been in support of Trump.

It was a bombshell. Interest in the term "fake news" spiked on

Google the day the story was published—and it stayed high for years, first as Trump's critics cited it as an explanation for the president-elect's victory, and then as Trump co-opted the term to denigrate the media at large.

The story concerned Facebook for reasons beyond the bad publicity. Even as the company's Communications staff had quibbled with Silverman's methodology, executives had demanded that News Feed's data scientists replicate it. Was it really true that lies were the platform's top election-related content?

A day later, the staffers came back with an answer: almost.

A quick and dirty review suggested that the data BuzzFeed was using had been slightly off, but the claim that partisan hoaxes were trouncing real news in Facebook's News Feed was unquestionably correct. Bullshit peddlers had a big advantage over legitimate publications—their material was invariably compelling and exclusive. While scores of mainstream news outlets had written rival stories about Clinton's leaked emails, for instance, none of them could compete with the headline "WikiLeaks CONFIRMS Hillary Sold Weapons to ISIS."

The finding rattled Facebook executives. Just days before, Zuckerberg had been saying that falsehoods on both sides would naturally cancel each other out, an argument that BuzzFeed's story and now their own research showed was wishful thinking at best.

There were so many things the company didn't know. Why had fake news started posting huge engagement numbers? Did the success reflect users' preferences, manipulation, or some flaw in Facebook's design? And was the problem worse on Facebook than on the rest of the internet or cable TV or at a Trump rally?

Addressing those questions was going to require Facebook to study and alter its platform in ways that it had considered inconceivable just months before. It would have to determine the truth and then alter its ecosystem in ways that favored it. The task would have been daunting under the best of circumstances—and the aftermath of a vicious presidential campaign that was beginning to focus on Russian election interference was hardly that.

With Zuckerberg's approval, Mosseri dispatched a team of his

News Feed staffers to begin quantifying the problem of fake news and coming up with potential solutions. To focus on the sensitive work in privacy, the team packed up their desks alongside the rest of the News Feed ranking staff and resettled in the closest thing they could find to a tucked-away corner in Building 20, the world's largest open-plan office.

Designed by Frank Gehry and completed just the year before, the space was mostly exposed structural supports and glass walls, a nod to Facebook's corporate embrace of connection and transparency. It hadn't been built for people to hide.

3

The prospect that Facebook's errors could have changed the out-come of the election and undermined democracy bothered Chris Cox more than any other top executive at the company. Cox, Face-book's chief product officer, was shaken in the week that followed Trump's victory, and he didn't save his feelings for his inner circle alone. A few days after the vote, an engineering executive at another company that he had unsuccessfully tried to poach emailed Cox to offer his services, as well as a blunt appraisal: Facebook had clearly screwed up, the executive wrote, and the platform was, in its current form, a threat to healthy public discourse.

"Good," Cox emailed in response to the man's offer, adding that he personally felt terrible.

Cox had long been concerned with the platform's societal effects, and particularly its potential role in stoking divisions. Back in 2015, he had surprised a candidate for an executive role with the scale of his ambitions. "I asked him the most important thing I could work on, and [Cox] basically said polarization, people feeling pushed apart," the executive recalled. "There was this real sense that Face-book could play a role in bringing them back together."

Facebook wouldn't be starting from scratch in the aftermath of the election, as it began to think more seriously about what it allowed on its platform. The company had established certain basic defenses. Primary responsibility for problems like spam and bulk data theft fell to the Site Integrity team; problems involving regular user mis-behavior were the province of a unit known as Community Operations. The need for such work reflected the growing importance of the platform to users, advertisers, and fraudsters around the globe.

Beyond Facebook's scale and prominence, there was another reason the platform needed more policing. Facebook was no longer the social network that Zuckerberg had famously built in his Harvard dorm room. Piece by piece, the product that employees called Blue had been replaced. The resulting product wasn't just more complex—it was more volatile.

Facebook's path to becoming the world's largest social media company was alternately paved with ambition and justifiable paranoia. As convinced as Zuckerberg was that Facebook was on a trajectory to change the course of the internet and society, he viewed any growing platform that allowed people to message, share, or broadcast content as an existential threat.

That belief dated back to 2004, in the days when the platform was run out of a rented house in Palo Alto. At the first sign that some other social media–like product might be catching on at college campuses, Zuckerberg would declare a "lockdown," during which everyone was expected to work near-nonstop until Facebook had finished building whatever features it considered necessary to blunt the threat.

Such do-what-it-takes scrambling became a point of pride. By 2011, when Facebook had a corporate campus, a red neon sign reading LOCKDOWN stood ready for the next emergency. A two-month sprint triggered by rumors of Google's plans to launch a social network produced Facebook's Groups and Events features, and Zuckerberg ended an internal talk on Google's threat by referencing a Roman orator who ended every speech with the words "*Carthago delenda est*"—Carthage must be destroyed.

Even after the sparsely adopted Google Plus had been vanquished, there was a reason the company could never ease up. Network effects made social media a winner-take-all game, one in which rival platforms were both a threat and a hindrance to the free flow of information.

"Mark viewed Snapchat like Ronald Reagan viewed the Berlin Wall," recalled a member of Facebook's Core Data Science team.

Facebook explicitly boasted that it wanted to build features that could accommodate every part of offline life. Zuckerberg had bought WhatsApp for $19 billion in 2014 after expressing fears about competition from text messaging, and he was pushing Facebook to compete with dating sites and peer-to-peer marketplaces. The CEO's demands for growth were steep. According to a later history of Facebook's metrics goals, Zuckerberg internally declared, in 2014, that Facebook should seek 10 percent annual usage growth *in perpetuity*. Responsibility for that infinite exponential expansion fell both on the elite Growth team and Engineering teams across the company.

As Zuckerberg explained to his chief financial officer in a 2012 email, he believed there were only "a finite number of different social mechanics to invent." Whenever a new one emerged, Zuckerberg wrote, the company would have to scramble to either copy it or acquire it. "The basic plan would be to buy these companies and leave their products running while over time incorporating the social dynamics they've invented into our core products," Zuckerberg wrote. "One way of looking at this is that what we're really buying is time."

This was the less fun and more paranoid interpretation of the company's well-worn motto "Move Fast and Break Things." As much as a cavalier exhortation to floor it on product development, the slogan reflected the need to worry about whatever was coming up in the rearview mirror. The Red Book, an internal company manifesto published shortly after the company's IPO, turned that fear into an inspirational slogan: "If we don't create the thing that kills Facebook, someone else will."

With every hot new competing product feature sparking an existential crisis, Facebook was bound to become a bit of a Frankenstein, hastily sewn together from rival platforms' parts. The company borrowed its reshare button from Twitter's retweets, launched live video following the success of Periscope, ripped off ephemeral video posts from Snapchat, and incorporated group video chats à la Houseparty. The blatant theft was viewed first as embarrassing, then craven, then comic. "Facebook Copied Snapchat a Fourth Time, and Now All Its Apps Look the Same," read a memorable *Recode* headline.

The need to adapt quickly meant code was regularly rough around the edges and did, in fact, "break things." A 2015 presentation by Ben Maurer, who oversaw the reliability of Facebook's infrastructure, noted that site failures tended to occur pretty much whenever Facebook engineers got near a keyboard. During one six-month period, the only two weeks where nothing major had gone wrong were "the week of Christmas and the week when employees are expected to write peer reviews for each other."

The engineers weren't incompetent—just applying often-cited company wisdom that "Done Is Better Than Perfect." Rather than slowing down, Maurer said, Facebook preferred to build new systems capable of minimizing the damage of sloppy work, creating firewalls to prevent failures from cascading, discarding neglected data before it piled up in server-crashing queues, and redesigning infrastructure so that it could be readily restored after inevitable blowups.

The same culture applied to product design, where bonuses and promotions were doled out to employees based on how many features they "shipped"—programming jargon for incorporating new code into an app. Conducted semiannually, these "Performance Summary Cycle" reviews incented employees to complete products within six months, even if it meant the finished product was only minimally viable and poorly documented. Engineers and data scientists described living with perpetual uncertainty about where user data was being collected and stored—a poorly labeled data table could be a redundant file or a critical component of an important product. Brian Boland, a longtime vice president in Facebook's Advertising and Partnerships divisions, recalled that a major data-sharing deal with Amazon once collapsed because Facebook couldn't meet the retailing giant's demand that it not mix Amazon's data with its own.

"Building things is way more fun than making things secure and safe," he said of the company's attitude. "Until there's a regulatory or press fire, you don't deal with it."

The constant stream of emergencies and shoddily built features led to dark jokes that Facebook was the world's oldest startup, with sloppiness its culture's most enduring feature. In 2021, when Mike Schroepfer, then chief technology officer, asked the company's engineers what their greatest frustration was, the response was overwhelming. "We perpetually need something to fail—often fucking spectacularly—to drive interest in fixing it, because we reward heroes more than we reward the people who prevent a need for heroism," read the top response in Workplace, which garnered around a thousand positive reactions and scores of comments recounting catastrophic-but-avoidable engineering failures.

Nowhere in the system was there much place for quality control. Instead of trying to restrict problem content, Facebook generally preferred to personalize users' feeds with whatever it thought they would want to see. Though taking a light touch on moderation had practical advantages—selling ads against content you don't review is a great business—Facebook came to treat it as a moral virtue, too. The company wasn't failing to supervise what users did—it was neutral.

Though the company had come to accept that it would need to do some policing, executives continued to suggest that the platform would largely regulate itself. In 2016, with the company facing pressure to moderate terrorism recruitment more aggressively, Sheryl Sandberg had told the World Economic Forum that the platform did what it could, but that the lasting solution to hate on Facebook was to drown it in positive messages.

"The best antidote to bad speech is good speech," she declared, telling the audience how German activists had rebuked a Neo-Nazi political party's Facebook page with "like attacks," swarming it with messages of tolerance.

Definitionally, the "counterspeech" Sandberg was describing didn't work on Facebook. However inspiring the concept, interacting with vile content would have triggered the platform to distribute the objectionable material to a wider audience.

The realities of the platform's mechanics notwithstanding, such faith in Facebook's goodness was often used to give Facebook's quest for growth the sheen of moral imperative. Such confidence reached its peak in an internal memo by Andrew "Boz" Bosworth, who had gone from being one of Mark Zuckerberg's TAs at Harvard to one of his most trusted deputies and confidants at Facebook. Titled "The Ugly," Bosworth wrote the memo in June 2016, two days after the murder of a Chicago man was inadvertently livestreamed on Facebook. Facing calls for the company to rethink its products, Bosworth was rallying the troops.

"We talk about the good and the bad of our work often. I want to talk about the ugly," the memo began. Connecting people created obvious good, he said—but doing so at Facebook's scale would produce harm, whether it was users bullying a peer to the point of suicide or using the platform to organize a terror attack.

That Facebook would inevitably lead to such tragedies was unfortunate, but it wasn't the Ugly. The Ugly, Boz wrote, was that the company believed in its mission of connecting people so deeply that it would sacrifice anything to carry it out.

"That's why all the work we do in growth is justified. All the questionable contact importing practices. All the subtle language that helps people stay searchable by friends. All of the work we do to bring more communication in. The work we will likely have to do in China some day. All of it," Bosworth wrote.

Facebook and Bosworth would both disavow his argument when *BuzzFeed News* published the memo two years later, but that didn't detract from his point. Facebook *had* misused people's contact lists to boost growth. It *had* written its privacy policies in ways that stymied and confused users who sought to make their profiles more private. And it *had* worked to build censorship tools as part of its effort to gain access to China. These things were ugly, and Facebook had done them consciously. Bosworth's memo wasn't a thought experiment. It was the code the company lived by.

Not every Facebook product feature was copied from others. In 2006, the U.S. patent office received a filing for "an automatically generated display that contains information relevant to a user about another user of a social network." Rather than forcing people to search through "disparate and disorganized" content for items of interest, the system would seek to generate a list of "relevant" information in a "preferred order."

The listed authors were "Zuckerberg et al." and the product was the News Feed.

The idea of showing users streams of activity wasn't entirely new—photo-sharing website Flickr and others had been experimenting with it—but the change was massive. Before, Facebook users would interact with the site mainly via notifications, pokes, or looking up friends' profiles. With the launch of the News Feed, users got a constantly updating stream of posts and status changes.

The shift came as a shock to what were Facebook's then 10 million users, who did not appreciate their activities being monitored and their once-static profiles mined for updated content. In the face of widespread complaints, Zuckerberg wrote a post reassuring users, "Nothing you do is being broadcast; rather, it is being shared with people who care about what you do—your friends." He titled it: "Calm down. Breathe. We hear you."

Hearing user complaints wasn't the same thing as listening to them. As Chris Cox would later note at a press event, News Feed was an instant success at boosting activity on the platform and connecting users. Engagement quickly doubled, and within two weeks of launch more than a million members had affiliated themselves with a single interest for the first time. The cause that had united so many people? A petition to eradicate the "stalkeresque" News Feed.

The opaque system that users revolted against was, in hindsight, remarkably simple. Content mostly appeared in reverse chronological order, with manual adjustments made to ensure that people saw both popular posts and a range of material. "In the beginning, News Feed ranking was turning knobs," Cox said.

Fiddling with dials worked well enough for a little while, but everyone's friend lists were growing and Facebook was introducing new features such as ads, pages, and interest groups. As entertainment, memes, and commerce began to compete with posts from friends in News Feed, Facebook needed to ensure that a user who had just logged on would see their best friend's engagement photos ahead of a cooking page's popular enchilada recipe.

The first effort at sorting, eventually branded "EdgeRank," was a simple formula that prioritized content according to three principal factors: a post's age, the amount of engagement it got, and the interconnection between user and poster. As an algorithm, it wasn't much—just a rough attempt to translate the questions "Is it new, popular, or from someone you care about?" into math.

There was no dark magic at play, but users again revolted against the idea of Facebook putting its thumb on what they saw. And, again, Facebook usage metrics jumped across the board.

The platform's recommendation systems were still in their infancy, but the dissonance between users' vocal disapproval and avid usage led to an inescapable conclusion inside the company: regular people's opinions about Facebook's mechanics were best ignored. Users screamed "stop," Facebook kept going, and everything would work out dandy.

By 2010, the company was looking to move beyond EdgeRank's crude formula to recommend content based on machine learning, a branch of artificial intelligence focused on training computers to design their own decision-making algorithms. Rather than programming Facebook's computers to rank content according to simple math, engineers would program them to analyze user behavior and design their own ranking formulas. What people saw would be the result of constant experimentation, the platform serving up whatever it predicted was most likely to generate a like from a user and evaluating its own results in real time.

Despite the growing complexity of its product and the collection of user data at a scale the world had never seen, Facebook still didn't know enough about its users to show them relevant ads. Brands loved the attention and buzz they could get from creating content on Facebook, but they hadn't found the company's paid offerings compelling. In May 2012, General Motors killed its entire Facebook advertising budget. A prominent digital advertising executive declared Facebook ads "fundamentally some of the worst performing ad units on the Web."

Fixing the problem would fall to a team run by Joaquin Quiño-nero Candela. A Spaniard who grew up in Morocco, Quiñonero was living in the UK and working on artificial intelligence at Microsoft in 2011 when friends scattered across Northern Africa began talking excitedly about social media–driven protests. The machine learning techniques he was using to optimize Bing search ads had clear applications to the social networks that people had used to overthrow four autocratic states and nearly topple several more. "I joined Facebook because of the Arab Spring," Quiñonero said.

Quiñonero found that the way Facebook built its products was nearly as revolutionary as their results. Invited by a friend to tour the Menlo Park campus, he was shocked to look over the shoulder of an engineer making a significant but unsupervised update to Facebook's code. Confirming how much faster the company moved than Microsoft, Quiñonero received a Facebook job offer a week later.

Quiñonero began working on ads, and his timing could hardly have been better. Advances in machine learning and raw computing speed allowed the platform to not only pigeonhole users into demographic niches ("single heterosexual woman in San Francisco, late twenties, interested in camping and salsa dancing") but to spot correlations between what they clicked on and then use that information to guess which ads they would find relevant. After beginning with near-random guesses on how to maximize the odds of a click, the system would learn from its hits and misses, refining its model for predicting which ads had the best shot at success. It was hardly omniscient—recommended ads were regularly inexplicable. But the bar for success in digital advertising was low: if 2 percent of users clicked on an ad, that was a triumph. With billions of ads served each day, algorithm tweaks that produced even modest gains could bring in tens or hundreds of millions of dollars in revenue. And Quiñonero's team found that it could churn out those alterations. "I told my team to go fast, to ship every week," he said.

The rapid pace made sense. The team's AI was improving not just revenue but how people felt about the platform. Better-targeted ads meant Facebook could make more money per user without increasing the ad load, and there wasn't all that much that could go wrong. When Facebook pitched denture cream to teenagers, nobody died.

Advertising was the beachhead for machine learning at Facebook, and soon everyone wanted a piece of the action. For product executives tasked with increasing the number of Facebook groups joined, friends added, and posts made, the appeal was obvious. If Quiñonero's techniques could increase how often users engaged with ads, they could increase how often users engaged with everything else on the platform.

Every team responsible for ranking or recommending content rushed to overhaul their systems as fast as they could, setting off an explosion in the complexity of Facebook's product. Employees found that the biggest gains often came not from deliberate initiatives but from simple futzing around. Rather than redesigning algorithms, which was slow, engineers were scoring big with quick and dirty machine learning experiments that amounted to throwing hundreds of variants of existing algorithms at the wall and seeing which versions stuck—which performed best with users. They wouldn't necessarily know why a variable mattered or how one algorithm outperformed another at, say, predicting the likelihood of commenting. But they could keep fiddling until the machine learning model produced an algorithm that statistically outperformed the existing one, and that was good enough.

It would be hard to conceive of an approach to building systems that more embodied the slogan "Move Fast and Break Things." Facebook wanted only more. Zuckerberg wooed Yann LeCun, a French computer scientist specializing in deep learning, meaning the construction of computer systems capable of processing information in ways inspired by human thinking. Already renowned for creating the foundational AI techniques that made facial recognition possible, LeCun was put in charge of a division that aimed to put Facebook at the vanguard of fundamental research into artificial intelligence.

Following his success with ads, Quiñonero was given an equally formidable task: pushing machine learning into the company's bloodstream as fast as possible. His initial staff of two dozen—the team responsible for building new core machine learning tools and making them available to other parts of the company—had grown in the three years since he'd been hired. But it was still nowhere near large enough to assist every product team that wanted machine learning help. The skills to build a model from scratch were too specialized for engineers to readily pick up, and you couldn't increase the supply of machine learning PhDs by throwing money around.

The solution was to build FB Learner, a sort of "paint by numbers" version of machine learning. It packaged techniques into a template that could be used by engineers who quite literally did not understand what they were doing. FB Learner did for machine learning inside Facebook what services like WordPress had once done for building websites, rendering the need to muck around with HTML or configure a server unnecessary. Rather than setting up a blog, however, the engineers in question were messing with the guts of what was rapidly becoming a preeminent global communications platform.

Many at Facebook were aware of the increasing concerns around AI outside the company's walls. Poorly designed algorithms meant to reward good healthcare penalized hospitals that treated sicker patients, and models purporting to quantify a parole candidate's risk of reoffending turned out to be biased in favor of keeping Black people in jail. But these issues seemed remote on a social network.

An avid user of FB Learner would later describe machine learning's mass diffusion inside Facebook as "giving rocket launchers to twenty-five-year-old engineers." But at the time, Quiñonero and the company spoke of it as a triumph.

"Engineers and teams, even with little expertise, can build and run experiments with ease and deploy AI-powered products to production faster than ever," Facebook announced in 2016, boasting that FB Learner was ingesting trillions of data points on user behavior every day and that engineers were running 500,000 experiments on them a month.

The sheer amount of data that Facebook collected—and ad-targeting results so good that users regularly suspected (wrongly) the company of eavesdropping on their offline conversations—gave rise to the claim that "Facebook knows everything about you."

That wasn't quite correct. The wonders of machine learning had obscured its limits. Facebook's recommendation systems worked by raw correlation between user behavior, not by identifying a user's tastes and interests and then serving content based on it. News Feed couldn't tell you whether you liked ice skating or dirt biking, hip-hop or K-pop, and it couldn't explain in human terms why one post appeared in your feed above another. Although this inexplicability was an obvious drawback, machine learning–based recommendation systems spoke to Zuckerberg's deep faith in data, code, and personalization. Freed from human limitation, error, and bias, Facebook's algorithms were capable, he believed, of unparalleled objectivity—and, perhaps more important, efficiency.

A separate strain of machine learning work was devoted to figuring out what content was actually in the posts Facebook recommended. Known as classifiers, these were AI systems trained to perform pattern recognition on vast data sets. Years before Facebook's creation, classifiers had proven themselves indispensable in the fight against spam, allowing email providers to move beyond simple keyword filters that sought to block mass emails about, say, "Vi@gra." By ingesting and comparing a huge collection of emails—some labeled as spam, some as not spam—a machine learning system could develop its own rubric for distinguishing between them. Once this classifier was "trained," it would be set loose, analyzing incoming email and predicting the probability that each message should be sent to an inbox, a junk folder, or straight to hell.

By the time machine learning experts began to arrive at Facebook, the list of questions that classifiers sought to answer had grown well past "Is it spam?," thanks in large part to people like LeCun. Zuckerberg was bullish on its future progress and its applications for Facebook. By 2016, he was predicting that classifiers would surpass human capacities of perception, recognition, and comprehension within the next five to ten years, allowing the company to shut down misbehavior and make huge leaps in connecting the world. That prediction would prove more than a little optimistic.

Even as techniques improved, data sets grew, and processing sped up, one drawback of machine learning persisted. The algorithms that the company produced stubbornly refused to explain themselves. Engineers could evaluate a classifier's success by testing it to see what percentage of its judgment calls were accurate (its "precision") and what portion of a thing it detected (its "recall"). But because the system was teaching itself how to identify something based on a logic of its own design, when it erred, there was no human-cognizable reason why.

Sometimes mistakes would seem nonsensical. Other times they would be systematic in ways that reflected human error. Early in Facebook's efforts to deploy a classifier to detect pornography, Arturo Bejar recalled, the system routinely tried to cull images of beds. Rather than learning to identify people screwing, the model had instead taught itself to recognize the furniture on which they most often did.

The problem had an easy fix: engineers simply needed to train the model with more PG-rated mattress scenes. It made for a good joke—as long as you didn't consider that the form of machine learning that the engineers had just screwed up was one of the most basic that Facebook was using. Similarly fundamental errors kept occurring, even as the company came to rely on far more advanced AI techniques to make far weightier and complex decisions than "porn/not porn." The company was going all in on AI, both to determine what people should see, and also to solve any problems that might arise.

There was no question that the computer science was dazzling and the gains concrete. But the speed, breadth, and scale of Facebook's adoption of machine learning came at the cost of comprehensibility. Why did Facebook's "Pages You Might Like" algorithm seem so focused on recommending certain topics? How had a video snippet from a computer animation about dental implants ended up being seen a hundred million times? And why did some news publishers consistently achieve virality when they just rewrote other outlets' stories?

Faced with these questions, Facebook's Communications team would note that the company's systems responded to people's behavior and that there was no accounting for taste. These were difficult points to refute. They also obscured an uncomfortable fact: Facebook was achieving its growth in ways it didn't fully understand.

Within five years of announcing that it was beginning to use machine learning to recommend content and target ads, Facebook's systems would rely so heavily on AI capable of training itself that, without the technology, Yann LeCun proudly declared, all that would be left of the company's products would be "dust."

Like most everyone on Facebook product teams, Joaquin Quiñonero had what he referred to as "the engineering mindset." Success was identifying a problem and then building something effective, powerful, and widely used to address it. "I obviously did not anticipate the unintended consequences of integrating simplified machine learning into a product," he said.

The arrival of machine learning in feed ranking was a big deal at the time, though few if any inside Facebook initially understood that the company was crossing a Rubicon. Facebook wasn't just changing what order people saw posts in—it was changing the entire dynamic of online social life.

Before ranked feed, friending someone meant enqueueing all their future posts for your review, an unappealing prospect for a casual acquaintance or a hyper-posting relative. Just as in offline life, there were only so many people you could stay in touch with, only so many groups you could join.

Ranking did away with that social carrying cost. Users could follow hundreds of accounts and pages without their account being overwhelmed; distant connections would ideally show up only when they posted something exceptionally engaging. The change made friending less significant and users stopped curating their friend groups—a sea change from the offline social world. With friend lists swelling, the platform became more public, encouraging people to create content that would be popular with a broad audience and to avoid posting more intimate things that they didn't wish to broadcast to casual acquaintances, or even people they didn't really know.

As the platform matured and its user base grew into the billions, Facebook scrambled to deploy the right metrics for success. From the start, the one most prized by senior leadership was Daily Average People, or DAP, the number of individual users who logged in on any given day. Once that metric became inured to Facebook's efforts to accelerate its growth, the company turned to sessions, the cumulative number of times human beings logged onto Facebook each day, as well as time spent and metrics for the production and consumption of specific types of content. Finally, there was a vast category of engagement metrics that ranged from fundamental (reshares) to arcane (interactions with birthday-related notifications).

One goal conspicuously missing from regular discussion was money. Even for employees responsible for ad targeting, "relevance" was supposed to trump dollars and cents. Among product staff, discussing the financial implications of Facebook's decisions was verboten, and conflicts between usage and profit were generally decided in usage's favor. "We don't build services to make money; we make money to build better services," Zuckerberg had written in a 2012 Red Book meant to encapsulate Facebook's revolutionary ethos. Placed on the desk of every new employee, the book was nominally created to celebrate the platform's reach to a billion users. But it came just a few months after Facebook's initial public offering. In the context of a massive and rapidly expanding market, the company's mission of making the world more open and connected could sometimes be hard to distinguish from the more craven pursuit of locking down market share.

Over time, as growth became harder to achieve, Facebook increasingly relied on one particular feature of its platform: virality. Zuckerberg's own love of the phenomenon went on display in 2014, when he took part in the Ice Bucket Challenge—dumping a bucket of ice water on his head to raise awareness and money for research to cure the degenerative neurological disease ALS and posting the video to Facebook, along with millions of others. The Ice Bucket Challenge raised more than $100 million for the disease's lead charity, the ALS Association. Zuckerberg would for years cite it as proof of the company's world-changing benefits.

To foster virality, Facebook changed its News Feed design to encourage people to click on the reshare button or follow a page when they viewed a post. Engineers altered Facebook's algorithm to increase how often users saw content reshared from people they didn't know. Facebook also began hunting for "friction"—anything that was slowing users down or limiting their activity—with the goal of eliminating it. One such change gave users the ability to create an unlimited number of pages. Another allowed them to cross-post the same material to multiple groups at once. A third prioritized recommending accounts that were likely to accept a friend request. Though the mechanics differed, all served to make Facebook faster.

Occasionally an employee would observe that the behaviors the company was encouraging looked a little odd. In a note titled "The Friending 1%," a data scientist noted that after years spent "optimizing friending volume," friending behavior was growing increasingly unequal. Just 0.5 percent of accounts were responsible for the majority of new connections formed, and those users were sending more than fifty requests per day. Was it possible that these promiscuous friend-makers weren't creating relationships of genuine value?

In response, colleagues mused about tweaking Facebook's "People You May Know" algorithm or restricting users from making more than one hundred friend requests on consecutive days. But nobody seemed to consider Facebook's lopsided friend growth to be an urgent problem.

The company's tools were becoming more complex, but its goals and methods weren't. Facebook wanted users to post, interact, and consume more content, and would change the platform in any way that made them do so, after brief tests. The company wanted growth, and it wasn't looking too closely at how it obtained it.

—

Before the 2016 election, as all these changes took root, much of Chris Cox's concern stemmed from what was happening with Facebook abroad. Married to a Thai woman who had directed a well-regarded documentary about a Bangkok sex worker, Cox was more attuned to Facebook's international users than most. (He is also a practicing Buddhist and a famously nice guy.)

Facebook had jumped headfirst into international expansion in 2008, crowdsourcing from users the translation of its site. Within two years, the company offered its services in seventy-five different languages. Hiring people who spoke those languages, however, was another story.

The company's overseas push unfortunately came just as Sheryl Sandberg was trying to impose cost controls ahead of Facebook's initial public offering, and she had ordered a headcount freeze at Community Operations, the team tasked with removing slurs, threats, fraud, and other illegal activity when it found them, generally via user reports.

"We were behind even when it started, and all these other regions were becoming hotspots," said Charlotte Willner, who headed the International Support team for nearly three years. "If I wanted to hire someone on Egyptian Arabic, I would have to not hire on Korean." Even in some of Facebook's largest overseas markets, the company was entirely reliant on Google Translate to police user content.

One day, Willner happened to read an NGO report documenting the use of Facebook to groom and arrange meetings with dozens of young girls who were then kidnapped and sold into sex slavery in Indonesia. Zuckerberg was working on his public speaking skills at the time and had asked employees to give him tough questions. So, at an all-hands meeting, Willner asked him why the company had allocated money for its first-ever TV commercial—a recently released ninety-second spot likening Facebook to chairs and other helpful structures—but no budget for a staffer to address its platform's known role in the abduction, rape, and occasional murder of Indonesian children.

Zuckerberg looked physically ill. He told Willner that he would need to look into the matter. Within the week, Willner's headcount request was funded.

She didn't blame Zuckerberg personally for the state of affairs. "There's no way Mark thought this was a tradeoff he wanted to make, but somewhere along this giant chain someone made that call," she said. "It's how large organizations work."

Willner's play for resources may have been a win, but it didn't amount to a wake-up call. By the time the general freeze in headcount was lifted, Willner said, the company was hopelessly behind in the markets where she believed Facebook had the highest likelihood of being misused. When she left Facebook in 2013, she had concluded that the company would never catch up.

The U.S. presidential election was still on the horizon in the summer of 2016 when Cox convened a regular working group of engineers, sociologists, economists, and data scientists. His concerns over polarization had been mounting, and a central topic of concern for his newfound group would be "filter bubbles."

It was a phenomenon that had been popularized years earlier in a TED Talk by Eli Pariser, the founder of the progressive advocacy group MoveOn.org. The concept was simple: as internet companies increasingly personalized the content they recommended to users, those people would be given a narrower view of the world. The Arab Spring was then unfolding, and Pariser demonstrated the idea of filter bubbles by describing what happened when he asked friends to Google "Egypt." One friend's results were packed with links to information about the ongoing revolt. A second friend's results omitted the uprising in favor of travel tips and pictures of pyramids. Pariser's 2011 talk kicked off a debate about whether such personalization might create "echo chambers" in which people were fed only information that conformed to their worldview.

In 2015, Facebook published a paper in the prestigious journal *Science* examining whether its News Feed recommendations exposed users to a more partisan diet of information than they would be likely to see on their own. The paper's verdict, based on a review of 10 million users' News Feeds, was that "individual choices more than algorithms limit exposure to attitude-challenging content." In a press release, Facebook plugged the research as debunking concerns about filter bubbles. A critical review in *Wired* raised questions about the methodology of what came to be known as "Facebook's 'it's not our fault' study," but the work nonetheless garnered thousands of citations in future research.

Equally concerning was an analysis shared within the working group about a particular strain of news publications on Facebook, specifically those catering to rabid partisans. Sites with generic names along the lines of 365USANews were scoring huge Facebook traffic with grammatically and factually incorrect headlines.

"It looked like crazy extremist sites were living off nutrients that Facebook was giving them. If Facebook cut them off, Google and organic traffic wouldn't support them," a data scientist involved in the work recalled. "This was the first real documentation of that."

As remarkable as the finding might be—amid an election season, at that—the team responsible for such sites wasn't being asked to make immediate changes. The work was simply for the company's own understanding. "It was relatively slow-moving, and the people involved were a little more academic," recalled a person involved by virtue of his position on the News Feed ranking team.

Even had the members of the group been geared for action, they would have faced a hurdle: Zuckerberg had personally ordered that the company move away from anything that looked like human curation.

The order stemmed from a controversy over Facebook's "Trending Topics," a regularly updated list of the hottest news stories on the site. Implemented in 2014 as a naked shot at competing with Google News and Twitter, Trending Topics saw Facebook use its automated systems to surface subjects that were surging in popularity. Members of a small curation team would then select a credible, representative news story to promote with a brief writeup. Some of the material was politics, some was entertainment, and all of it was confined to a small box in the upper right-hand corner of the site. The product was a minor one, fundamentally unconnected to Facebook's core News Feed ranking.

The feature's sleepy status changed in May 2016, though, when Michael Nuñez, a reporter for the tech site *Gizmodo,* dropped a series of articles about its opaque mechanics. In a first article, Nuñez interviewed people who edited Trending Topics, revealing a group of underemployed Ivy League media types hired via a staffing agency and working out of the basement of Facebook's New York headquarters (two would turn out to be Nuñez's former roommates). "Nobody really knows much about how it works—and the company isn't telling," the journalist wrote.

The company, which provided only a cursory comment for the story, would come to regret not playing ball. While the article created buzz in media circles, it was nothing compared with Nuñez's follow-up a week later, titled "Former Facebook Workers: We Routinely Suppressed Conservative News." The headline's explosive claim of political bias largely relied on one former contractor who had kept a log of topics popular with conservatives that he believed his colleagues had snubbed, such as allegations about IRS commissioner Lois Lerner and any stories about the controversial conservative YouTube comic Steven Crowder.

Nowhere in Nuñez's article was there any claim that the company itself had asked anyone to put their thumb on the scale against conservatives. The only demand from Facebook management was that the team avoid highlighting stories about Facebook itself. "In other words, Facebook's news section operates like a traditional newsroom, reflecting the biases of its workers and the institutional imperatives of the corporation," Nuñez wrote. "Imposing human editorial values onto the lists of topics an algorithm spits out is by no means a bad thing—but it is in stark contrast to the company's claims that the trending module simply lists 'topics that have recently become popular on Facebook.'"

Even if the substance of the story boiled down to "Facebook engaged in modest human curation," the specter of intentional bias opened a can of worms at the very moment that Trump's demonization of the media was becoming central to his campaign.

Katie Harbath was in the Philippines at the time the story dropped, learning about it when outraged posts by Republican friends started popping up in her News Feed. She handed her phone to a Facebook comms staffer and told them the company was about to become a centerpiece of conservative allegations of bias in media.

She was right. The task of tamping down the outcry fell to Molly Cutler, a lawyer and trusted deputy to Sheryl Sandberg. Part of Cutler's work involved conducting an internal review, through which she found that the attention paid to conservative and liberal news outlets in Trending Topics was "virtually identical." Trump had been the most widely discussed topic for both.

But that wasn't enough for Zuckerberg. The company hurriedly organized a meeting between the CEO and "conservative thought leaders" including Glenn Beck, Tucker Carlson, and Barry Bennett, a Trump campaign aide, in which Zuckerberg pledged to vigilantly defend the platform's neutrality. "I wanted to hear their concerns personally and have an open conversation about how we can build trust," Zuckerberg posted on Facebook after the meeting.

Within a few months, Facebook laid off the entire Trending Topics team, sending a security guard to escort them out of the building. A newsroom announcement said that the company had always hoped to make Trending Topics fully automated, and henceforth it would be. If a story topped Facebook's metrics for viral news, it would top Trending Topics.

The effects of the switch were not subtle. Freed from the shackles of human judgment, Facebook's code began recommending users check out the commemoration of "National Go Topless Day," a false story alleging that Megyn Kelly had been sacked by Fox News, and an only-too-accurate story titled "Man Films Himself Having Sex with a McChicken Sandwich."

Setting aside the feelings of McDonald's social media team, there were reasons to doubt that the engagement on that final story reflected the public's genuine interest in sandwich-screwing: much of the engagement was apparently coming from people wishing they'd never seen such accursed content. Still, Zuckerberg preferred it this way. Perceptions of Facebook's neutrality were paramount; dubious and distasteful was better than biased.

"Zuckerberg said anything that had a human in the loop we had to get rid of as much as possible," the member of the early polarization team recalled.

Among the early victims of this approach was the company's only tool to combat hoaxes. For more than a decade, Facebook had avoided removing even the most obvious bullshit, which was less a principled stance and more the only possible option for the startup. "We were a bunch of college students in a room," said Dave Willner, Charlotte Willner's husband and the guy who wrote Facebook's first content standards. "We were radically unequipped and unqualified to decide the correct history of the world."

But as the company started churning out billions of dollars in annual profit, there were, at least, resources to consider the problem of fake information. In early 2015, the company had announced that it had found a way to combat hoaxes without doing fact-checking— that is, without judging truthfulness itself. It would simply suppress content that users disproportionately reported as false.

Nobody was so naive as to think that this couldn't get contentious, or that the feature wouldn't be abused. In a conversation with Adam Mosseri, one engineer asked how the company would deal, for example, with hoax "debunkings" of manmade global warming, which were popular on the American right. Mosseri acknowledged that climate change would be tricky but said that was not cause to stop: "You're choosing the hardest case—most of them won't be that hard."

Facebook publicly revealed its anti-hoax work to little fanfare in an announcement that accurately noted that users reliably reported false news. What it omitted was that users also reported as false any news story they didn't like, regardless of its accuracy.

To stem a flood of false positives, Facebook engineers devised a workaround: a "whitelist" of trusted publishers. Such safe lists are common in digital advertising, allowing jewelers to buy preauthorized ads on a host of reputable bridal websites, for example, while excluding domains like www.wedddings.com. Facebook's whitelisting was pretty much the same: they compiled a generously large list of recognized news sites whose stories would be treated as above reproach.

The solution was inelegant, and it could disadvantage obscure publishers specializing in factual but controversial reporting. Nonetheless, it effectively diminished the success of false viral news on Facebook. That is, until the company faced accusations of bias surrounding Trending Topics. Then Facebook preemptively turned it off.

The disabling of Facebook's defense against hoaxes was part of the reason fake news surged in the fall of 2016. With the presidential election just months away, Facebook's leadership had just taken its foot off the brakes.

"All this was a direct consequence of the media focus on the trending stories stuff," the News Feed ranking staffer noted. "It led to a terrible blind spot."

4

The Federal Bureau of Investigation wanted to meet James Barnes and his lawyers in the parking lot of a Washington, DC, Holiday Inn.

A black town car pulled up with two men in the front seat, who told Barnes to sit in the middle, between his two Facebook-provided lawyers. Then they drove him to the nondescript office building that was home to Robert Mueller's investigation of Russian interference in the 2016 election. Upstairs, the men launched into their questioning. Were there any Russians in the San Antonio office, they wanted to know.

"You guys have got to be kidding me," Barnes replied.

"Okay, but you know we had to ask," an agent responded.

Mueller had been running a special counsel investigation into possible Russian interference in Trump's election since May 2017, as the country reeled from reports of contacts between Trump's circle and Russian officials while it tried to understand how someone like Trump could have made it to the White House.

Facebook got its first warning that Russia was attempting to interfere in the upcoming election in 2015, along with the other major tech companies. But the focus was on what Russians might do *to* social media platforms, not the risk that they'd become popular *on* them. Reporters, including Adrian Chen at the *New York Times Magazine,* had written about a troll farm called the "Internet Research Agency" based in St. Petersburg, but it never occurred to anyone at Facebook that a band of malicious foreign users might run a mass-scale manipulation effort on a social network of 2 billion users.

The company wasn't alone. The government had been focused on the prospect of Russians attempting to mess with the election results, not sway voters. "The idea of applying influence operations to the internet and social media of 2016, people weren't focused on that," said Nathaniel Gleicher, a former director of cybersecurity policy for the National Security Agency, who joined Facebook as head of Cybersecurity Policy at the beginning of 2018.

In the wake of the election, people both inside and outside Facebook were asking a lot of questions that would have seemed implausible before. A team led by Alex Stamos, Facebook's chief security officer, began looking for networks of Russian users who had been playing in American politics, and they found them. Separate from the Russians' large-scale hack and leak operations—they targeted the Democratic National Committee and then the emails of Hillary Clinton's campaign manager, John Podesta—they had invested in building up partisan accounts and pages on Facebook that had pumped political bile into the presidential race. Most of the material was hostile to Clinton, but the networks of fake accounts had dabbled in anti-Trump propaganda, too. The problem was industry-wide. Similar postmortems were going on at Twitter and YouTube, with similar results.

The company's first response to Stamos's findings was dismay. The second was to wonder what else it still didn't know. Mueller's investigation was just getting underway, and the press was full of stories about contacts between Trump associates and Russian agents. Was it possible that people in Trump's orbit had been privy to Russia's social media manipulation efforts? The company undertook an exhaustive review, scrutinizing what Trump associates had searched for on Facebook, who they messaged, and what content they viewed, looking for potential evidence of collaboration or foreknowledge. They found none.

Facebook's review left the company in the same boat as much of Washington and the media—keenly aware that social media had played a crucial role in the 2016 election, but not clear about how.

The story got worse with the discovery that the Russians had spent $100,000 on Facebook election ads, some of which they had paid for in rubles. Facebook looked at the spending and saw that it fit a pattern used by garden-variety viral marketers. The Russians had created pages with names like Blacktivist and Secured Borders, then bought what were called "Page Like" ads in an effort to cheaply acquire followers. The ads are exactly what they sounded like—their only call to action was to like a page, thereby ensuring that the viewer would get served future posts that the page owner wasn't paying to promote. Once a page got big enough, the ad buys trailed off, and the page gathered an audience with "organic" distribution. Page Like ads were a way for the Russians to buy their way to mass scale. "Page Like ads were the on-ramp to the misinformation superhighway," as one Facebook director put it.

Facebook was taking a beating every time a new revelation about Russian activities emerged, and Sheryl Sandberg had had enough. She wanted to make clear that the company wasn't hiding anything. To do that, the company needed to show it wasn't afraid to turn over rocks. She approached the team doing the forensics on the Russians' operations with an odd request: What was the ugliest, most shocking thing the company could reveal about interference on the platform? Once that came out—by Facebook's own hand, no less—the company, she believed, would be able to put the scandal behind it. Researchers told her to look at "reach," a term drawn from advertising that counts the total number of people who simply saw a particular piece of content at least once. It would be vastly greater than, for instance, the number of posts Russians had made or how many people had engaged with them. The company would release the "reach" statistic at a congressional hearing on social media manipulation, one that would also be attended by representatives from YouTube and Twitter.

On the eve of the hearing, in the fall of 2017, reporters reviewed the prepared testimony of Facebook legal counsel Colin Stretch and his counterparts at Twitter and Google. Twitter had acknowledged finding 131,000 Russian tweets. Google had disclosed that "more than 1,000" Russian videos had been uploaded. For its part, Facebook was coming to the table with a number in the nine figures. The comparison wasn't apples to oranges so much as an apple to a school bus. As a portion of the trillions of total content views over the course of the 2016 election, it was trivial—a rounding error—but it sure sounded like a lot:

"Russian Influence Reached 126 Million Through Facebook Alone," a *New York Times* headline blared.

Throughout the controversies around Russia and fake news, Zuckerberg maintained an appearance of calm, leaving the cleanup to his deputies. The company says he followed the progress closely, and the CEO had other things to do, such as completing his annual personal challenge that he posted on Facebook. These challenges sometimes tied in with Facebook's own struggles: When Zuckerberg was facing questions about whether he was mature enough to be a CEO, he publicly resolved to don a tie every day. When Facebook was looking to get into China, he had taken up the challenge of learning Mandarin. Zuckerberg's 2017 goal, chosen in the wake of a wildly divisive election, was to visit every state. "My work is about connecting the world and giving everyone a voice," he wrote. "I want to personally hear more of those voices this year."

Then came Cambridge Analytica.

On March 17, 2018, front-page stories at the *New York Times* and the *Guardian* unveiled the sensational claims of Christopher Wylie, a pink-haired former employee of the British digital strategy firm. The company had clear links to the Mercer family and billionaire Trump donors, and Wylie said he believed the firm had been making overtures to Russians.

The real story was more boring than the daily churn of stories at the time might have led people to believe. Starting in 2010, Facebook had let outside developers access data from the profiles of Facebook users when they used the developer's products. This information included not just a user's own activity but that of their friends, including posts, follows, and likes. As the company's awful data privacy practices came to light in the years that followed, Facebook curtailed "open graph" access—but the previously accessed data was out of its hands. Cambridge Analytica acquired just such data and put it to work building "psychographic profiles," a nebulously defined term that implied the capacity for deep, almost subconscious levels of psychological manipulation via "behavioral micro targeting." There's no sign the Trump campaign ever made use of the information, or that it would have been much help if it did—the data Cambridge Analytica had was partial and two years old, a lifetime for ad-targeting purposes.

The deepest irony of all, according to Barnes and Harbath, was that Facebook itself would gladly have provided the slicing and dicing of demographic data that Cambridge Analytica was allegedly doing with purloined data. The platform had long provided major advertisers with the ability to target users with data the platform didn't make available to small businesses—sales reps would create such bespoke audiences using off-menu data at a customer's request. Part of Barnes's job in San Antonio had been to create such audiences to order for Trump's campaign, just as other Facebook employees did for clients like Coca-Cola. One former executive involved in the company's internal review of the Cambridge Analytica debacle put it this way: "If the campaign had wanted Facebook to build a custom collection of psychological profiles, they wouldn't need to have gotten it from a fucking British hack academic."

Although there were ample grounds for Facebook to suspect that Cambridge Analytica had been selling snake oil, uncertainty about what data Cambridge Analytica possessed and how it had been used made it hard for the company to credibly argue that the scandal wasn't a big deal. Combining technology, secretive billionaires, and mercenary sleaze—one senior Cambridge Analytica executive got caught discussing the possibility of entrapping political leaders overseas with prostitutes—the scandal captured the conspiratorial nature of the thinking at the time. It tied all the strings together into a comprehensive, if not altogether credible, theory for how social media had been abused, voters had been manipulated, and Trump had been elected—all with the help of Facebook.

Regardless of whether Cambridge Analytica had any influence

on American politics, the effects were immediately felt inside Facebook. In the months before the scandal broke, the head of Facebook's ad growth team, Rob Goldman, had already been pushing to shut down Facebook's relationship with major third-party data brokers who provided targeting information for advertisers, arguing that it posed a privacy risk to Facebook and its users. The dispute rose to the level of Sandberg. She heard out Goldman but decided that the data about users' offline lives was too valuable to give up. When the Cambridge Analytica story broke, she reversed herself the next day.

As Facebook's senior leadership was busy trying to deflect one incoming scandal after the next, its Engineering teams were busy trying to head off new ones. Of particular concern to Carlos Gomez-Uribe, a Mexican machine learning specialist, was what might charitably be called problem content. Put another way, it was the junk that was now routinely clogging users' feeds.

Gomez-Uribe had been recruited by Facebook in January 2017 to lead News Feed recommendations. He had proved his mettle at Google and then at Netflix, which he joined in 2010 just as the company was shifting from mail-order DVD rentals to streaming video, a transition powered in part by the company's famed recommendation system. Using billions of data points about users' rentals and ratings, the company offered personalized movie picks that differentiated it from competitors.

But Gomez-Uribe, hired as a statistician, considered the company's approach far too basic. Netflix had the world's most comprehensive data set about people's taste in movies, and it was still making recommendations based on a five-star scale that a critic for the *New York Daily News* had invented in 1929. When he told the manager in charge of the system that he thought he could do better, the manager gave him a chance to prove himself or die trying.

In less than a week, Gomez-Uribe produced a rough prototype for a recommendation system that promised to outperform the one Netflix's engineers had spent years building.

The new model that emerged from that effort—heavily based on machine learning techniques—scoured every bit of behavioral data the company had in order to guess not just what titles a user would like but the set of options they wanted right this second. Recommendations varied according to the time of day, a user's proclivity to binge-watch, and even geographic trends. If a major hurricane was bearing down on Florida, Netflix would cue up *The Perfect Storm* and *Sharknado.*

Gomez-Uribe had turned down job offers from Facebook during the early years of its machine learning push, but he reconsidered in the aftermath of the 2016 election, proposing himself for a job cleaning up News Feed ranking. The director-level position he was offered would be a substantial step down for a Netflix VP with a staff of a hundred, but the issues seemed interesting and Gomez-Uribe considered Facebook in need of help.

Patience not being among Gomez-Uribe's strengths, his acceptance came with one stipulation: he wanted an assurance that he could meaningfully change the product, like he had at Netflix. The streaming service had certainly cared about metrics—boosting user engagement through recommendations, for instance. Subscribers who watched more TV were more likely, after all, to keep paying for the service. But the goal there had been to create long-term value, and that wasn't an easy thing to quantify. It had been okay, for example, for his team to rework Netflix's suggestions in a way that reduced daily logins if there was a valid justification. Would Facebook allow that?

At a meeting with Zuckerberg, Gomez-Uribe tried to figure out how sincere Facebook was about overhauling its algorithms. How would Zuckerberg respond if cleaning up News Feed came at the expense of the company's traditional growth metrics?

"If you find good approaches that significantly improve integrity at the expense of engagement, that's fine—we'll launch them," Gomez-Uribe recalled Zuckerberg saying. "Because I'm also assuming that the Growth team will do their job and growth will continue."

Reassured, Gomez-Uribe accepted the job. His conviction that

Facebook's recommendation systems needed fixing only grew more intense after he had a chance to see from the inside how the company built them.

Facebook's ranking work was sloppy—there was no other way to put it. The company altered its recommendation systems on the basis of A/B tests that ran for just a few weeks, months less than the period Netflix considered necessary to observe longer-term shifts in user behavior. Facebook was more cavalier in shaping a global communications and broadcasting platform than Netflix was about deciding to steer users toward *The Great British Bake Off*. (A company spokeswoman disputed Gomez-Uribe's recollection of hasty and inadequate testing of ranking changes, calling the company's experiments rigorous and focused on improving user experience.)

Beyond the "Done Is Better Than Perfect" approach to experimentation, Gomez-Uribe also questioned what Facebook was measuring. Maybe the platform needed to rethink how it tallied engagement, in the same way Netflix had rethought what it meant for a video to be a good match.

There were precedents in Facebook's history for the kind of systematic rethinking Gomez-Uribe was imagining. Its first approach to ranking—based on likes—had been crude, and within a few years it became clear that users were liking their friends' posts not because they actually appreciated them but as a simple marker that told their friend, "Yes, I saw this." With the value of likes debased, the company needed some new user behavior to maximize. It began to prioritize time spent on the platform.

But changing metrics can change behaviors, and refocusing on time spent led to Facebook encouraging users to consume a lot of video, to the detriment of other activities. This came with its own disadvantages; so Facebook backtracked a little and gave more weight to content that produced a mix of emojis, comments, and likes.

The process was humbling for the company. Nobody had predicted the devaluation of likes, or that showing users more videos would decrease overall content production on the platform, even if in hindsight both made sense. In recognition of how much Facebook didn't know, Mosseri took to exhorting the News Feed team to always be on the lookout for "one-way doors" in the product, meaning alterations that would change it in ways that couldn't be readily rolled back if the platform surprised its overseers.

"It was admirable, and I still think about it all the time," said one former senior Engineering manager. "But if you go through too many doors, you can't get back to the first one—two-way doors become one way over time." In other words, some changes aren't so easily undone.

The company was making dozens of changes to ranking each quarter, with every new decision layered upon those that had come before. The cumulative weight of all the tweaking tended to discourage big alterations. Thousands of engineers had spent years optimizing the platform based on agreed-upon metrics for success. Changing anything foundational would be tantamount to throwing away years of hard-won gains.

That's precisely what Gomez-Uribe, heavily influenced by his time at Netflix, wanted to do: throw the whole thing out. He thought back to how the streaming service had learned to care about how *long* users spent on each piece of content. If people quit watching a TV show after three minutes, that was a worse sign than if they never watched it at all.

Gomez-Uribe's team began playing around with a new system that would reward only substantive engagement—seeing something and immediately clicking reshare would be worth less than, say, having a post open for a minute and then leaving a comment. It encouraged the kind of interaction they sought, but as a way to address integrity problems the approach fell on its face—the amount of time a user spent with a piece of content didn't reliably correlate with whether it was accurate and substantive.

A second approach was more promising, though controversial. Facebook's most intense users, Gomez-Uribe posited, might not be good for the platform. Data analysis showed that there was a small but significant portion of users that seemed to be spending astonishing amounts of time on Facebook. Sometimes the numbers suggested bad faith. How could an account that logged in twenty hours a day not be either automated or run by people working in shifts?

Gomez-Uribe's team hadn't been tasked with working on Russian interference, but one of his subordinates noted something unusual: some of the most hyperactive accounts seemed to go entirely dark on certain days of the year. Their downtime, it turned out, corresponded with a list of public holidays in the Russian Federation.

"They respect holidays in Russia?" he recalled thinking. "Are we all this fucking stupid?"

But users didn't have to be foreign trolls to promote problem posts. An analysis by Gomez-Uribe's team showed that a class of Facebook power users tended to favor edgier content, and they were more prone to extreme partisanship. They were also, hour to hour, more prolific—they liked, commented, and reshared vastly more content than the average user. These accounts were outliers, but because Facebook recommended content based on aggregate engagement signals, they had an outsized effect on recommendations. If Facebook was a democracy, it was one in which everyone could vote whenever they liked and as frequently as they wished.

After working at Netflix and Google—both of which capped the amount of influence any individual user could have on recommendations—Facebook's approach struck Gomez-Uribe as bizarre. Netflix would never build an algorithm that gave someone who watched eighty hours of TV every week ten times the influence on its recommendations as someone who watched eight. Netflix had nothing against binge-watching, but the idea that all views were equal was foundationally distortive.

So Gomez-Uribe's team proposed a big but simple fix. Facebook would rebuild the News Feed algorithm to limit the influence of hyperactive users. The platform would still take their engagement into account, but the more they engaged, the less weight Facebook's recommendations would give to each additional action. This would require changing the math of how the engagement of every page, group, and comment got scored. It would take around twenty people to rebuild the system, Gomez-Uribe expected, and, once completed, News Feed would have to be retrained in a speed run of years' worth of optimization efforts. It would be no small undertaking, precisely the kind of ambitious project Gomez-Uribe had been assured he could pursue.

Despite the work required, the idea immediately appealed to a lot of senior product executives who had worked at other major tech companies. Among those who backed it was Quiñonero, the company's top applied machine learning engineer. "There's a law of the universe that says 90 percent of the clicks have to come from fewer than 10 percent of the users, and that gives disproportionate influence to a small number of power sharers," Quiñonero said. "That's a design decision, and the only way to address it is to deeply rethink the metrics themselves."

But the merits of the change didn't seem so obvious to News Feed executives Jon Hegeman and Lars Backstrom, both of whom had spent virtually their entire careers steeped in Facebook's culture of growth and optimization. If some users happened to be extremely enthusiastic about sharing content and interacting with other users, why would Facebook want to mute them?

Gomez-Uribe's request was refused. So he did pretty much what he had done at Netflix, which was do it himself until management saw the light.

The lack of resources meant that Gomez-Uribe was forced to rein in his ambitions for an overhaul. With the help of volunteers from Facebook's Core Data Science team, Gomez-Uribe's staff worked up a quick and dirty version of his original plan. Rather than rebuilding News Feed to give quieter users more say, they would just slap a final equation onto the existing algorithm's final result for recommending news stories. By dampening the distribution of content disproportionately popular with hyperactive users, the platform would be lowering the volume of people desperate to make Facebook their personal megaphone. This stripped-down approach wasn't ideal, but it worked.

Because hyperactive users tended to be more partisan and more inclined to share misinformation, hate speech, and clickbait, the intervention produced integrity gains almost across the board. An analysis of how the intervention would affect the distribution of polarized content in the United States showed it would hit far-right and far-left outlets hard—and slightly boost the distribution of mainstream news publishers.

Given how much bigger the American conservative publisher ecosystem was than the left's, there was no question that views of content popular with Republicans would drop more. But that finding didn't hold across the globe—in Venezuela, for example, the change ended up affecting Facebook's dominant leftists more than their opponents.

There was no question the approach, nicknamed "Sparing Sharing," was novel and interesting: Facebook filed a patent for it in September 2017. In a nod to how obviously beneficial the results seemed, Gomez-Uribe began referring to the U-shaped graph of the impact as "the happy face." With this effort, altering News Feed to prevent loudmouths from dominating the distribution of news stories, Gomez-Uribe managed to win a sign-off from both Hegeman and Mosseri in News Feed—what was not to like?

Plenty, according to Facebook's Public Policy team. Conservative publishers were already accusing the company of targeting their content via biased fact-checks, overly aggressive content moderation, and secret suppression efforts. Even though there was nothing inherently political about limiting users' ability to turbocharge content distribution through sheer persistence, there was no question that conservative and liberal content fared differently on Facebook. Conservative voices outperformed—and got into trouble more.

The asymmetry had sparked heated debate inside the company. Did American conservatives misbehave more or was Facebook's enforcement just biased? Had conservatives flocked to Facebook for information because mainstream outlets were stockpiled with raging leftists? Were liberal constituencies fractured in ways that made achieving mass scale harder, or had their online organizers become complacent during the Obama administration? Might American conservative rhetoric rely more heavily on anger and fear, and did Facebook disproportionately reward that? (In Facebook's socially liberal culture, this last question tended to be asked in a whisper.)

Whatever the explanation, conservative content's outperformance on Facebook was a fact the company had to reckon with. Even if Sparing Sharing applied the same rubric to publications on the right and left, the net loss to conservatives was greater. And that was going to be awkward.

Had Facebook's algorithm originally been built in the way that Gomez-Uribe proposed, none of this would have been a problem—nobody was going to call for congressional hearings because some News Feed engineers had limited the effect a single account could have on recommendations. But changing the system now would upset an entire industry of Facebook-native publishers that had evolved to harness the engagement of hyperactive users. What was Facebook's responsibility to the constituencies of the online society it had shaped?

A profound moral and philosophical question was at stake here, one with global implications. So Facebook checked in with its top advocate in Washington, Joel Kaplan, whose formal title was head of Global Public Policy.

That a company committed to neutrality would run questions of platform design and mechanics by its chief lobbyist was not standard. At Twitter and Google, sensitive decisions touching on content moderation or platform design ran through teams of product policy specialists, who reported in turn to the company's legal department. At Facebook, though, such calls were overseen by somebody who had been deputy chief of staff in the George W. Bush White House and then lobbied for a financially troubled power company out of Dallas.

At one meeting about Sparing Sharing, Kaplan and Elliot Schrage—Facebook's head of Public Policy and Communications, there as a representative of Zuckerberg and Sandberg—grilled Gomez-Uribe about his proposed intervention. The anticipated damage to conservative media outlets' traffic was an obvious backdrop to the discussion. But the explicit point of concern was that Gomez-Uribe's proposal was premised on the idea that there was something problematic about extremely heavy Facebook use. What if, say, a troop of Girl Scouts became superusers to promote the sale of cookies? (Kaplan said in a later interview that he had no recollection of citing the Girl Scouts, but acknowledged concern that the ranking change would harm publishers who had cultivated unusually enthusiastic followers on Facebook.)

The strenuous internal opposition didn't kill Sparing Sharing—but it did mean Gomez-Uribe's plan would need the blessing of a higher authority.

Booking a block of time on Zuckerberg's calendar took weeks and the assistance of Mosseri as well as News Feed ranking executives, who remained supportive of the initiative, if a little more tentative following the spat with Public Policy. When the time came, in late 2017, Gomez-Uribe went with data in hand to argue a case he considered obvious. His proposed change would give the moderate majority of Facebook's users more of a voice in what news spread across the platform and significantly reduce the spread of stories that ended up being flagged as false by fact-checkers.

"I still thought that presumably the people in charge would ask me to do more of this kind of stuff," he recalled.

Zuckerberg heard out Gomez-Uribe and Kaplan for ten minutes before rendering his judgment.

"Do it, but cut the weighting by 80 percent," Zuckerberg told Gomez-Uribe. "And don't bring me something like this again."

Gomez-Uribe's plan had unknowingly run up against a question of ideology, but not of the political variety. Facebook had been built to produce the maximum total engagement. The idea that the platform needed to be protected against the excesses of its most enthusiastic users simply wasn't welcome.

Gomez-Uribe left the meeting angry. With the penalty for hyper-shared links only one-fifth the size of what his team had considered effective, the arrival of Sparing Sharing wasn't going to meaningfully change the platform.

"Carlos built the levers pretty quickly with minimal support, and he got the watered-down 'Let's make our people happy' version rather than the 'We actually want to do this' version," a colleague recalled.

Mosseri encouraged him to look on the bright side, congratulating him on getting a groundbreaking launch approved at any strength. A partial victory was a victory nonetheless.

Fuck you. How is that a success? Gomez-Uribe thought.

He had pushed for a revolution—and gotten a consolation prize instead.

Gomez-Uribe dusted himself off. In collaboration with a researcher focused on polarization, he plunged right into another sensitive area. Facebook data could easily distinguish between pages that catered to a narrow segment of users versus those that appealed to a diverse audience. Why not alter ranking to favor pages that seemed to help users from different camps get along?

After the company's unenthusiastic response to Sparing Sharing, there was no question that Gomez-Uribe's team was playing around with something above its pay grade. They tried to draw as little attention as possible to their plan. They would slip their experiment into the stream of algorithm changes that Facebook was constantly testing. If the effort showed promise, then the data they generated would provide ammunition to fight for it.

The plan almost worked. With the experiment on the verge of launch—the only thing left to do was for another team to flip a single bit of code from 0 to 1, Gomez-Uribe recalled—a member of Facebook's News Feed marketing team got wind of the effort and ratted it out to colleagues on Facebook's Public Policy and Communications teams. Gomez-Uribe was immediately besieged with hostile questions about the effort. Wasn't suppressing polarizing content a form of social engineering? How would Facebook publicly explain what he was doing?

It had just been an experiment, but it was looking to gather data that Facebook actively didn't want. With other divisions of the company up in arms, the leadership of the News Feed ranking team stepped in and blocked the experiment.

"Carlos was a crusader," Michael McNally, head of Connections Integrity, said. "He wanted to save the world."

The incident marked the early death of Facebook's most direct effort to combat polarization—and the end of Gomez-Uribe's limited patience. In May 2018, less than a year and a half after Facebook hired him to reimagine how to address content problems through ranking, he quit and went to Stanford, pursuing an interest in statistics and theoretical physics.

As the Cambridge Analytica scandal continued to suck up public attention, Zuckerberg didn't measure the fallout in damning *New York Times* editorials or inflammatory Rachel Maddow clips. What bothered him was metrics.

The truth was that, as rough as the Russian influence stories had been for the company, they hadn't seriously shaken Facebook's benchmark for user trust, measured through a daily survey of users known as CAU. Short for "Cares About You," the statistic was gathered by polling thousands of users with a single question: "Do you believe Facebook cares about you?" The casual-sounding nature of the question belied its importance. CAU's precise wording had been intensely workshopped across the world and found to be the optimal gauge of user sentiment about the company.

CAU had gradually declined over the years, as Facebook grew bigger and more corporate, but it had only once meaningfully moved in a way that showed up in the day-to-day numbers—after Facebook chose to put its messaging features into a separate app. (This transgression took users many months to forgive.)

Now, in the wake of the Cambridge Analytica scandal, CAU was tanking, taking its worst hit of all time. The problem had Zuckerberg's full attention. A sustained hit to CAU, he believed, could do permanent damage to Facebook.

Zuckerberg was obsessed with the idea of "breaking through" on Cambridge Analytica. Whenever Facebook got in a tough spot, the company's response was to wait a few days and then announce something dramatic and new. Rather than ignoring or responding to criticism, the approach was based on supplanting it with buzz, and it had generally worked. But nothing was breaking through Cambridge Analytica. Not the announcement of internal probes. Not the release of pioneering new privacy controls. Not pledges to spend billions of dollars on content moderation.

Publicly, Zuckerberg and Facebook weren't going to dismiss Cambridge Analytica's significance. In speeches and posts, the CEO said the company was remorseful and committed to self-examination. Inside the company, however, he took a different tack.

At a meeting in June 2018, Zuckerberg declared himself a "wartime CEO," a reference to the writings of Ben Horowitz, cofounder, alongside Facebook board member Marc Andreessen, of famed venture capital firm A16Z. According to Horowitz's book *The Hard Thing About Hard Things*, "Peacetime CEO knows that proper protocol leads to winning. Wartime CEO violates protocol in order to win." Other successful "wartime CEO" traits, per Horowitz's book, included paranoia, the purposeful use of profanity, and yelling.

Zuckerberg apparently heeded the advice. He told employees at a company-wide town hall that the coverage of Cambridge Analytica was "bullshit," and he started becoming visibly angry during meetings about how to change the conversation to anything but the scandal. "I could come up with better ideas if I spent ten minutes thinking about it," Zuckerberg fumed in one such meeting with executives.

For solutions, Zuckerberg turned to Naomi Gleit, then-head of Social Good. Gleit was one of Zuckerberg's longest-serving deputies, part of the small group of executives who had gained the CEO's trust through long tenures. She was known as a troubleshooter along the lines of *Pulp Fiction*'s Mr. Wolf, an efficient manager who could be counted on for flawless, step-by-step execution in dire circumstances. If Gleit was put in charge of a job, it was a likely bet that it was both important and that someone else had already fucked it up.

Gleit's division held a prominent position within Facebook's Growth team. She oversaw the company's upbeat, societally positive work, like the launch, in 2015, of on-Facebook fundraisers, which the company pitched as geared toward encouraging spontaneous, small-dollar generosity and helping users connect with causes they cared about. Omitted from public statements about the project was any reference to how handling donations would give the company access to users' credit card data, an invaluable resource for a company interested in internet commerce. Internal dashboards used to gauge the success of Facebook fundraisers included the number of credit cards processed as a central metric of success, alongside metrics such as total dollars raised for charity.

The Russia fallout and Cambridge Analytica scandal didn't just piss off average users. In June 2018, Harbath, as director of Global Elections Policy, hosted a group of liberal advocacy organizations at Facebook's DC headquarters. They were all angry to a degree that went beyond what their normal liaisons with the company could handle.

The immediate focus of their rage was Cambridge Analytica, but that, touching on Facebook's historically cavalier attitudes about user data, was also wrapped up with foreign election interference, misinformation, and inadequate moderation. How these admitted failures were interconnected was unclear. But when a Planned Parenthood representative declared that Facebook was killing people, Harbath knew not to argue the point.

"This was a time when a lot of stuff was getting jumbled together," she recalled. "My job was to be human and let people vent."

As soon as the liberals left, Harbath, along with Kaplan, met with a different set of visitors: Kevin McCarthy, the Republican House Majority Leader, and Brad Parscale, who had risen to become Trump's campaign manager. They, too, were upset with Facebook, though for different reasons. Parscale was convinced that the company maintained a secret list of phrases that it used to downrank conservatives who questioned climate change and other elements of liberal dogma (it did not). But what really seemed to bother him was that the company still hadn't thanked him for publicly crediting Facebook with Trump's 2016 victory. Parscale belabored the point to an extent that McCarthy snapped at him: "What do you want, Brad? For them to send you some flowers?"

McCarthy also had grievances. He was upset that Google's search results had briefly listed the California Republican Party as Nazi-affiliated earlier that month. The error had nothing to do with Facebook—it resulted from someone briefly vandalizing a Wikipedia page referenced by Google without detection—but McCarthy seemed to take it as evidence that all of Silicon Valley was out to get Republicans. Harbath and Kaplan both did their best to persuade him otherwise, but McCarthy and Parscale left unconvinced. (Parscale would later say that he did not recall the details of the meeting, but that he was tired of "lying and reacting to their woke employees," and that he believed anti-conservative censorship by the company likely cost Republicans the 2020 election.)

Such distrust was hard to dispel given that Facebook's staff was heavily liberal, with 87 percent of their political donations going to Democrats. Already facing an onslaught of illegitimate anti-conservative allegations of bias, the company was worried about the possibility of real ones. Kaplan became a lead internal watchdog, vetting product change proposals for bias or anything that could be alleged as such. Similarly, Kaplan held the reins on individual content moderation decisions—Monika Bickert, the company's head of Content Policy, reported to him. Everyone, of course, reported to Sandberg when there was a particularly important call to make, and she deferred to Zuckerberg on anything really touchy. While the CEO did sometimes intervene on content moderation and policy questions, he tended to consider them annoying distractions. In practice, Kaplan often had the final say on questions of ranking, rules, and enforcement—a sideline to his primary job of keeping Washington's powerful happy, or at least mollified.

As Kaplan's role inside Facebook became more widely understood outside Facebook, liberals began to mirror conservatives in alleging political favoritism. But although he did sometimes intervene on behalf of conservatives, not even his critics inside Facebook considered him a firebrand.

The true conflict of interest wasn't that the company had put a Republican lobbyist in charge of adjudicating its rules and opining on its mechanics, but that it had turned over those duties to a lobbyist of any stripe. Facebook's billions of users would be governed by someone whose job was to look out not for them but for Facebook.

5

When it came to cleaning up users' feeds, some quality issues ought to be easy pickings. That was clear to Michael McNally, who joined Facebook in April 2017 following thirteen years at Google, where he had worked on training early machine learning models to fight spam.

It would be difficult to overstate how urgent the problem of spam had been back at the turn of the twenty-first century, when bulk messages were poised to ruin email, and maybe communication on the internet more broadly. Spam grew exponentially at a time when the filters to stop it were making only incremental gains. Congress passed a law—the CAN-SPAM Act—and prosecutors sent a few people to jail for running spam operations, but it didn't ultimately fix anything. The machine learning work that people like McNally worked on did.

For humans, what qualified as spam was intuitive and immediately recognizable. For computers, not so much. A big part of the work, therefore, was figuring out what objective signals humans tended to produce when they ran into it. Google was already at the forefront of studying this when McNally arrived. In one particular feat of ingenuity, a Google engineer named Simon Tong had come up with a behavioral method for porn detection.

Tong's approach didn't work by evaluating images—back in those days, artificial intelligence for computer vision was crude by contemporary standards. All it did was look at where users' cursors moved when they received a bulk email. With regular emails, there would be no pattern. But with porn, a recipient's mouse would immediately begin hovering over the image like a pervy Irish setter. The response was near-involuntary, a momentary twitch that reflected what humans do when we are unexpectedly confronted by a naked fellow human—we look.

The bad guys had their patterns, too. For not just email but ad fraud and bogus reviews, McNally and his colleagues used a combination of automation, data analysis, and common sense to distinguish between malicious users and normal ones. The single biggest constant was that the bad guys tended to be a little too excited about using the company's products. Google wanted people to send lots of emails, click on lots of ads, and write lots of reviews—but only up to a point. Performing an action too much warranted suspicion if not outright hostility. The most dangerous users were the ones that looked, from the standpoint of simple usage, like the best ones.

"Things that were hyperactive and exceptional were untrustworthy," McNally said. Sometimes the misbehavior was automated, but not always. One internal Google product, called "Zip It," looked at the time distribution of a user's five-star ratings to battle attempts at manipulating the order of shopping listings. Displaying too much enthusiasm too fast about too many products was a marker of bad intent. When someone wrote twenty different five-star product reviews in a single day, the appropriate response was to disregard all of them.

Google's success at searching for the signature features of bad behavior produced services and search results that bested rival providers. Aggravations like spam and ad fraud would never be solved outright—there would always be bad actors and mistakes. But as long as Google stayed on top of training its detection models, the system by and large worked.

Facebook hired McNally to use the same detection and machine learning techniques to engage in its newfound battle with misinformation. Initially given a team of twenty staffers, McNally arrived both older than most Facebook executives and more jaded. After nearly two decades dealing with scammers in the bowels of the internet, he was not someone inclined to "assume good intent." That disinclination was exactly why the company needed him.

"Technology has unintended consequences, and the road to hell is paved with good intentions," McNally said in an interview years later. "Zuckerberg, Chris Cox, and others were admitting that something was wrong and that it could be made better. I didn't come to Facebook as an admirer, I came to try to be of service."

Along with disabusing others of the corporate mandate to think charitably, McNally also had to adjust to Facebook's approach to experimentation. At Google, any meaningful change to its search algorithm required a thorough survey of the fallout. Whatever the proposal was, the staffers pushing for it would draw up a chart for their superiors on the Search Quality team demonstrating its effects and why they thought the altered results of the new model were an improvement.

Facebook sometimes performed that level of evaluation, but it was too expensive and time-consuming to be routine. Instead, the worthiness of an algorithm change was judged on how it altered around a dozen core metrics, such as the length of users' sessions, how much content they produced, and how much they interacted with other users.

As with their counterparts at Google, the team proposing a change generally had a theory about why a tweak would improve the platform. For example, there's an intuitive logic to the idea that new users need to find their friends to make the platform fun. Would prioritizing fresh accounts in Facebook's People You May Know recommendations speed up the process and get new users off on a better foot?

With a plausible idea in hand, the team responsible for friend growth would test variations of the idea over the course of a few weeks on tiny fractions of Facebook's user base in minor markets, trying out versions until they identified whichever seemed to best increase friend growth. As long as nothing the team did damaged any other metric that Facebook cared about, a rejiggering of People

You May Know recommendations would be a strong candidate for launch.

It seemed like an intuitive approach, but, by McNally's thinking, there was a layer of analysis missing. There certainly was nothing wrong with Facebook rolling out the welcome wagon for new users, but did the results make sense in the real world? Were the new connections being formed genuine-seeming, or were they all with young women with the glossy good looks of stock photos? Did these new friends actually behave as if they truly knew each other?

At Facebook, he realized, nobody was responsible for looking under the hood. "They'd trust the metrics without diving into the individual cases," McNally said. "It was part of the 'Move Fast' thing. You'd have hundreds of launches every year that were only driven by bottom-line metrics."

Something else worried McNally. Facebook's goal metrics tended to be calculated in averages.

"It is a common phenomenon in statistics that the average is volatile, so certain pathologies could fall straight out of the geometry of the goal metrics," McNally said. In his own reserved, mathematically minded way, he was calling Facebook's most hallowed metrics crap. Making decisions based on metrics alone, without carefully studying the effects on actual humans, was reckless. But doing it based on *average metrics* was flat-out stupid. An average could rise because you did something that was broadly good for users, or it could go up because normal people were using the platform a tiny bit less and a small number of trolls were using Facebook way more.

Everyone at Facebook understood this concept—it's the difference between median and mean, a topic that is generally taught in middle school. But, in the interest of expediency, Facebook's core metrics were all based on aggregate usage. It was as if a biologist was measuring the strength of an ecosystem based on raw biomass, failing to distinguish between healthy growth and a toxic algae bloom.

Where Carlos Gomez-Uribe leveled blunt criticism directly at the executives who oversaw News Feed, McNally worked within the system. He had been enthusiastic about Gomez-Uribe's attempts to dial back News Feed's preferential treatment of hyperactive users and disappointed to see the effort stall, its effect cut by 80 percent at Zuckerberg's urging. But McNally's pragmatism mirrored that of the systems he tried to build. Within the parameters set by Facebook, his team would undertake the most effective steps it could.

A demonstration of that approach came with Facebook's early misinformation work. By 2017, Facebook had abandoned its conviction that, thanks to the wisdom of crowds, users would simply suss out falsehoods on their own and avoid spreading them. The revelations around the 2016 election had quickly given the lie to that line of thought. But Facebook remained unwilling to take on the job of determining truth itself, so it set up partnerships with outside fact-checking organizations.

Their journalistic efforts helped Facebook identify dodgy stories and then levy penalties against other stories emerging from the same domains. But these measures had only a limited effect. By the time Facebook tackled fake news, the Macedonians and their imitators had moved long past hobbyist status. They now maintained hundreds of largely throwaway sites with names like "The Denver Guardian." When one got targeted, the entity behind it could simply switch to another, promoting the content through networks of Facebook pages.

"At Facebook, the awareness that there were adversaries gaming the engine came in late," McNally said. To drive fake news purveyors off the platform, Facebook would have to make posting bullshit stories unprofitable on a consistent basis. And, as with McNally's previous work fighting spam and ad fraud, the first step was figuring out how the behavior of fake news publishers differed from that of legitimate outlets.

One distinguishing feature was the shamelessness of fake news publishers' efforts to draw attention. Along with bad information, their pages invariably featured clickbait (sensationalist headlines) and engagement bait (direct appeals for users to interact with content, thereby spreading it further).

Facebook already frowned on those hype techniques as a little spammy, but truth be told it didn't really do much about them. How much damage could a viral "Share this if you support the troops" post cause?

When such engagement-boosting chum was being used in the service of a fake news operation, the answer, it turned out, was quite a lot. And so McNally's team began working to dial up the company's enforcement against what had previously been treated as a petty misdemeanor. If Facebook could target hyperbolic language and efforts to artificially prod users into engaging, perhaps that would make the company's fake news problem more manageable.

To conduct such a crackdown, Facebook needed to build systems to detect bad posts. Engineers fed a machine learning system examples of sensationalistic and engagement-baiting posts, training it to distinguish between that material and more restrained content. The resulting algorithm predicted the likelihood that a post was bait on a scale of zero to one.

This technique, called supervised learning, was technically complex but conceptually straightforward. The goal was to reliably recognize all the bad content it saw without any false positives. Such automated perfection was impossible, of course, and tough decisions arose when determining how and whether to use a tool that would lead to inevitable errors. There was always that tradeoff between precision (how often a classifier erred) and recall (what percentage of the targeted content it actually caught).

Whether Facebook valued precision or recall more depended on whether it was more concerned about false negatives or false positives. If Facebook was contemplating the automated deletion of fake accounts, it needed to be more than 99 percent sure that a user was bogus before pulling the trigger, even if that level of precision meant leaving millions of fake accounts untouched. In other circumstances, recall mattered far more: a classifier that surfaced posts displaying potential suicidal intent should err on the side of caution, flagging content for human review even if it had a low likelihood of being actionable.

Most integrity problems fell between these two extremes. Facebook didn't want people seeing hate speech on its platform, but it wasn't willing to take down or manually review every post a classifier flagged as potentially offensive. Here, the company needed to hedge its bets.

The solution was downranking. Rather than deleting or reviewing a questionable post, Facebook could simply put its thumb on the scale of the News Feed to ensure that it reached fewer people, suppressing its score in the rankings that determined what showed up when in a user's feed. If the penalty was big enough, the questionable post would either be omitted entirely or appear so far down in the queue that a user was unlikely to ever reach it.

Downranking's advantage was that Facebook didn't have to decide whether the content was acceptable or not. A user's closest friends might still see the post, but for everyone else the demoted content would merely be replaced by whatever item was next in line. It also lowered the cost of imprecision. Since nothing got deleted, nobody got upset. But Facebook still didn't want to suppress content on shaky grounds. Nobody wanted the platform to serve as an automated morality police, burying any post with a bit of spicy language or a hint of skin. A perfectly safe platform would be perfectly boring.

How to calibrate the mixture of automated content removals and demotions was therefore both a practical and a philosophical matter. If you believed that bad content posed a serious danger and thought Facebook was responsible for what content it recommended, you likely favored a relatively heavy hand. But if you viewed Facebook as a neutral platform that provided users with personalized content, you would consider that approach needlessly strict, maybe even unethical. Users' News Feeds reflected their personal preferences and tastes, and Facebook should intervene in their expression as little as possible.

"Harm comes from situations where the viewer doesn't see something that they really wanted to see," stated one internal research note summing up the minimalist position and arguing that ranking changes that denied users access to maximally engaging content were failing to "respect" their judgment.

This debate bubbled up from time to time on Workplace, though absolutism from employees on either side was rare. The maximalists didn't think Facebook should enforce against all potentially unwholesome content, and the minimalists didn't think Facebook needed to honor white supremacists' appetite for racist memes. Facebook wasn't responsible for everything its users posted and consumed, but it wasn't the free-for-all of the open internet either.

Wherever one might land on this spectrum, there wasn't much question where Zuckerberg and most of Facebook's senior leadership fell. Despite efforts from some quarters to portray Facebook as a censorship-happy tyrant, the CEO always seemed more concerned about overdoing moderation. Facebook should govern its own platform as little as possible, removing only unambiguously bad content and requiring a preponderance of evidence for even modest content suppression. Those hurdles meant that integrity interventions happened largely on the margins—later research showed that News Feed tweaks downranking graphic violence, low-quality health information, and misinformation altered only a tiny percentage of users' sessions in any fashion, and even then, users didn't report that their feeds were less worth their time. Gomez-Uribe's Sparing Sharing demotion, in its weakened form, affected just one in a thousand.

A cynic could fairly note that such principled restraint had the added benefit of requiring less work and producing more engagement. But people who spoke with the CEO generally considered his position to reflect ideology more than profit or expediency. If you thought of users' Facebook feeds as reflecting their wishes—and Zuckerberg did—then you were suspicious of efforts to interfere with them.

Whatever their source, these leanings all but ruled out the most straightforward approach to combating misinformation: to simply downrank sensationalism and engagement bait more frequently and harder than they currently were, sacrificing precision to increase recall. Cranking up the dial in this way was off the table.

Facebook's mandate to respect users' preferences posed another challenge. According to the metrics the platform used, misinformation was what people *wanted*. Every metric that Facebook used showed that people liked and shared stories with sensationalistic and misleading headlines.

McNally suspected the metrics were obscuring the reality of the situation. His team set out to demonstrate that this wasn't actually true. What they found was that, even though users routinely engaged with bait content, they agreed in surveys that such material was of low value to them. When informed that they had shared false content, they experienced regret. And they generally considered fact-checks to contain useful information.

So, in solidly institutionalist fashion, McNally set up weekly meetings with Adam Mosseri and Jon Hegeman to make the case that, when it came to bait content and fake news, engagement metrics simply couldn't be relied upon to reflect what users actually valued.

His effort paid off. In May 2017, Facebook announced it would start demoting clickbait, not because it was misleading, sensational, or scammy, but because users told them that they didn't like things that were misleading, sensational, or spammy. Official neutrality, it turned out, was the key that could unlock an entire body of work.

"We try to find out what are annoying experiences, and we combat things independently of misinformation," McNally said of the approach.

The strategy was to ding hoax publishers for their nuisance behaviors, not for publishing fake news. Fake news sites tended to have a ton of ads, and people didn't like ads, providing justification for downranking. Such sites also tended to launch pop-ups, and everyone hated those. The fake news sites on average took a moment longer to load than more legitimate outlets, and nobody likes waiting.

Absent the problem of fake news, there was no chance that Facebook would have prioritized these initiatives, many of which were at least somewhat redundant. But there was a game being played here. To avoid having to adjudicate the truth, the company would punish hoax publishers for everything except the fact that their news stories weren't true.

Ranking wasn't the only lever available to McNally's team. They made fact-check labels that encroached on the physical space allotted to a false post, thereby depriving it of attention as a user scrolled.

And in an effort to give Facebook's fact-checkers a bit of a head start on viral false news, they built a classifier that screened comments for common expressions of disbelief, prioritizing posts with high ratios of skeptical comments.

Other social media platforms, like YouTube and Twitter, were adopting similar tactics, but Facebook arguably moved the fastest and was certainly the most transparent. Dozens of different initiatives launched over the course of 2017.

McNally had become adept at clearing internal hurdles, but sometimes the threats to misinformation work came from outside the company. One aggressive approach to fighting misinformation was to simply segregate news from users' feeds, placing it all in a separate tab that would function as a bit of a quarantine. Where once Facebook had pushed to expand news distribution in an effort to fend off Twitter, now it was trying to cut back. As with most major design changes, Facebook tested the separated news tab in a handful of smaller markets, including Bolivia, Cambodia, and Serbia.

The test, carried out in late 2017, succeeded in both tanking news consumption on the platform and infuriating publishers. Citing the plummeting traffic of a legitimate Bolivian news outlet that relied on Facebook, the *New York Times* accused Facebook of turning the country into "a guinea pig in the company's continual quest to re-invent itself."

Amid withering criticism from old-line and upstart digital publishers alike, Facebook called off the experiment. The irony was close to the surface. Facebook was being blamed by news publishers for degrading the entire news ecosystem. But when the company took a step that would have reduced its role in news distribution, publishers screamed.

"What the outside world wants of Facebook is contradictory and not satisfiable," McNally said.

Perhaps the news tab idea could have been explained better, or maybe it just wasn't the right approach. But every time a well-intentioned proposal of that sort blew up in the company's face, the people working on misinformation lost a bit of ground. In the absence of a coherent, consistent set of demands from the outside world, Facebook would always fall back on the logic of maximizing its own usage metrics.

"If something is not going to play well when it hits mainstream media, they might hesitate when doing it," McNally said. "Other times we were told to take smaller steps and see if anybody notices. The errors were always on the side of doing less."

Although McNally had some success in pushing the company to look beyond what users engaged with for evidence of what they wanted, there were limits. Nowhere did his team run into them more clearly than in its effort to create a new metric known as "Broad Trust."

For years, the company's Media Partnerships team had been trying to figure out ways to tilt Facebook's scales toward major respected publishers and away from the fly-by-night digital news sources that regularly outperformed them. The Partnerships team had proposed simply giving mainstream outlets a leg up—upranking them in feeds, for instance. That idea was shot down immediately. Facebook was no more eager to adjudicate quality than it was to adjudicate fact.

McNally saw an opportunity to get into similar territory in a way that didn't require the company to pick favorites. Rather than privileging what it considered reliable outlets over random upstarts, the platform could give an advantage to sources that users themselves trusted.

As much as people might be drawn to content from dubious publishers on Facebook, they rarely sought out the material off the platform, or even knew where it came from. If a user noticed that a news story a friend had shared was from a hitherto-unknown outlet like "The Boston Inquirer"—information that Facebook's interface hardly emphasized—they became more skeptical and less likely to share it. As it stood, Facebook wasn't taking such preference for known reliable sources into account.

So McNally's researchers began working on a metric to establish "Broad Trust," using surveys. One might have suspected the whole effort would be doomed by partisanship, with registered Democrats dismissing stories from Fox News and Republicans treating

MSNBC's stories as pure lies. But that wasn't how things worked, at least not in 2017. American Facebook users overwhelmingly rated both outlets as legitimate sources of news. In contrast, sites like "Ending the Fed," one of the most notorious publishers of fake news ahead of the 2016 election, scored poorly on trust, even with conservative users. McNally's proposal was to give outlets that had earned Broad Trust a leg up on the ones that hadn't.

As a way to fight misinformation and sensationalistic content, Broad Trust unquestionably worked. People were still free to consume and share content from whatever news site they wanted. But News Feed itself would become less likely to amplify articles from a news site that had been created the previous week.

Facebook approved the incorporation of Broad Trust into News Feed ranking in late 2017, with Zuckerberg personally announcing its arrival. "There's too much sensationalism, misinformation, and polarization in the world today. Social media enables people to spread information faster than ever before, and if we don't specifically tackle these problems, then we end up amplifying them," Zuckerberg wrote. "That's why it's important that News Feed promotes high quality news that helps build a sense of common ground."

Broad Trust was publicly a winner. Behind the scenes, however, it hadn't gone down well. Like Sparing Sharing, its effects weren't politically balanced; the ecosystem of partisan digital publishers was significantly larger on the right. Despite Zuckerberg's professed support for Broad Trust, and even though users themselves seemed to consistently support the idea behind it, Zuckerberg had decided—following vigorous lobbying by Joel Kaplan—that the strength of the ranking change's impact should be heavily watered down, just like Carlos Gomez-Uribe's Sparing Sharing.

There was good reason to consider that decision ominous: Broad Trust had little potential to give Facebook a black eye. Its effects were broadly appealing, and the only injured parties—fly-by-night news outlets—weren't a particularly powerful or beloved constituency. But even with the stakes low and the benefits significant, the company had still balked.

Facebook had supported McNally's work to target the absolute worst stuff—foreign publishers of outright lies. But they reined him in as soon as he tried to target the incentives that gave rise to that industry in the first place.

"For people who wanted to fix Facebook, polarization was the poster child of 'Let's do some good in the world,'" McNally said. "The verdict came back that Facebook's goal was not to do that work."

McNally got the message. Still determined to work within the system, he focused his efforts on doing things that the company would support. Over the course of a little more than a year, McNally's team shipped more than 140 separate ranking changes meant to strike at the business model of publishers of blatantly fake news. Many amounted to no more than paper cuts, but together they made an impact.

The cumulative weight of the changes brought the worst offenders to their knees. The change was pronounced enough to be readily visible outside the company. After a little more than a year of work, social media analytics company NewsWhip pronounced the old-guard hoax publishers pretty much dead.

Whatever the frustrations, this was a major success. When the ranking team had begun its work, there had been no question that Facebook was feeding its users overtly false information at a rate that vastly outstripped any other form of media. This was no longer the case (even though the company would be raked over the coals for spreading "fake news" for years to come).

Ironically, Facebook was in a poor position to boast about that success. With Zuckerberg having insisted throughout that fake news accounted for only a trivial portion of content, Facebook couldn't celebrate that it might be on the path of making the claim true.

Even as Facebook was effectively cleaning up its most obvious stumbles in the aftermath of the 2016 election, the platform was introducing new initiatives—and with them, new complexities.

At the beginning of 2018, Zuckerberg made a big announcement. Amid growing public concern about the effects of social media on mental health, Facebook, with the help of academics at the world's top universities, had taken a hard look at its products and found that passively consuming content wasn't good for people. To be beneficial, social media had to strengthen people's social ties.

"Based on this, we're making a major change to how we build Facebook," Zuckerberg declared in a Facebook post. "I'm changing the goal I give our product teams from focusing on helping you find relevant content to helping you have more meaningful social interactions."

A new metric, "Meaningful Social Interactions"—or MSI—was born.

The change would come at the expense of how much time people spent on Facebook, Zuckerberg said. But he was more concerned with "bringing people closer together" than quantifying their daily usage.

It all sounded great, and company user research had indeed shown that interacting with friends through Facebook was better for people than scrolling zombielike through News Feed. But there was a more prosaic explanation for the move: users weren't commenting enough.

Facebook had long sought to maximize "time spent." Though Zuckerberg's 2014 mandate of perpetual 10 percent growth was a long-term mathematical impossibility—there was, in the end, only a limited number of humans on earth with only a limited number of hours in the day—it was also an example of Facebook doing what it did best: setting a bold but easily measured goal, and then incenting product teams across the company to take a run at it.

At first they had succeeded in meeting Zuckerberg's lofty target, so much so that they raised it. The goal became constant annual-time-spent growth of 12 percent.

Then came what Facebook did worst.

"Nobody was really thinking much about the bigger picture," recounted a senior data scientist on the company's Analytics team, on the history of Facebook's goals. "In early 2017 engagement metrics started to dive . . . and nobody noticed." It was a company-wide screwup. Everyone had been chasing metrics like a pack of kindergartners chasing a soccer ball. Meanwhile, the data scientist said, "the system was tanking."

The decline was dangerous. Commenting and reactions were an essential component of Facebook's perpetual motion machine: without engagement, people stopped posting and reposting free content. And without free content, Facebook died.

By the time the company realized what had been happening right under its nose, circumstances were dire.

The plunging metrics kicked off a "sprint that never really figured out why metrics declined," the data scientist recounted. Part of it was likely a result of the turn to video consumption that setting goals for maximizing "time spent" encouraged. It also stemmed from users increasingly consuming content from large publishers, which fit into "an existing belief that FB was no longer focused on your friends," that data scientist added. "Having too much video and other public content was a problem."

News Feed was due for an overhaul. Changing how engagement was calculated would be a huge deal at Facebook—it represented an about-face from the way the company had measured its own success since its earliest days. But in November 2017—just as the company would normally be starting to wind down big projects for the year—Zuckerberg ordered News Feed to stanch the engagement losses. They had thirty days. (In a statement, the company denied that the process had been rushed, saying that the new algorithm had been built over eighteen months—but internal documents were unequivocal. The work began shortly before Thanksgiving and ended before the end of the year.)

A host of executives, including Mosseri and Hegeman, began trying to work out how to reweight the algorithm. "Meaningful Social Interactions" was a term of art; it did not yet have a definition. So they began with the company's existing metric for interaction, a raw count of inter-user actions known as U2U, and then tried to refine what kinds of actions were most important—most socially meaningful. Comments seemed more meaningful than emoji reactions, and emojis seemed more meaningful than likes. Comments and reshares fit in there somewhere, too. But how to weight these actions' relative importance was not yet clear.

Given the rushed timeframe of its development, MSI could hardly be a calibrated measure of what brought meaningful connection to users' lives. It included no effort at sentiment analysis, meaning it gave equal value to a heartfelt bereavement note and a declaration of intention to piss on the departed's grave. What mattered was not the content of the message but the fact of the comment itself. The company had already added a host of reaction emojis beyond the basic "like," to include "love," "haha," "wow," "sad," and "angry." By Facebook's new metric, when someone died, Facebook did not care if you chose a heart or an angry face, as long as you clicked on something.

Over a few scant weeks, the platform completely upended the worth it assigned to user behaviors. Under the new MSI system, a reshare was worth up to thirty times as much as a like. Comments were worth fifteen times as much, and emoji responses worth five. Engagement from friends a user interacted with most frequently got a little extra boost, as user experience research confirmed that people did care more about engagement coming from friends they interacted with more. But testing was rushed. Facebook urgently wanted to elicit specific user behaviors, so it simply changed News Feed to favor the content that would produce them.

"These guys were flying by the seat of their pants," recalled one former executive involved in the project.

Simply changing what content made it into people's feeds wasn't the end goal, of course. The hope was that, in response to the changes, some users would begin producing content that earned the reactions, shares, and comments that the new system valued. "Mimicry"—Facebook users' well-documented pattern of duplicating one another's behavior—would take over from there. User experience research confirmed intuition that people did care more about engagement coming from their friends, validating the newfound approach.

Facebook had just upended how it valued the activity of 2 billion people, on the fly. The metric was still rough, but its fundamentals were in place. News Feed would now prioritize whatever content generated the highest amount of "MSI given" and favor content producers who brought about outsized quantities of "MSI received." Special deference would be given to posts expected to generate "downstream MSI," the term for the avalanche of follow-on interactions garnered by repeatedly reshared viral posts.

This all made sense even if it was a little loose, the company said—it considered people's engagement to be a good proxy for how much value they derived from a post.

The Civic and News Feed Integrity teams weren't present for the birth of the metric. But multiple members of both teams recalled having had the same response when they first learned of MSI's new engagement weightings: *it was going to make people fight.* Facebook's good intent may have been genuine, but the idea that turbocharging comments, reshares, and emojis would have unpleasant effects was pretty obvious to people who had, for instance, worked on Macedonian troll farms, sensationalism, and hateful content.

Hyperbolic headlines and outrage bait were already well-recognized digital publishing tactics, on and off Facebook. They traveled well, getting reshared in long chains. Giving a boost to content that galvanized reshares was going to add an exponential component to the already-healthy rate at which such problem content spread. At a time when the company was trying to address purveyors of misinformation, hyperpartisanship, and hate speech, it had just made their tactics *more* effective.

Multiple leaders inside Facebook's Integrity team raised concerns about MSI with Hegeman, who acknowledged the problem and committed to trying to fine-tune MSI later. But adopting MSI was a done deal, he said—Zuckerberg's orders.

Even non-Integrity staffers recognized the risk. When a Growth team product manager asked if the change meant News Feed would favor more controversial content, the manager of the team responsible for the work acknowledged it very well could.

"The News Feed Integrity team is working very hard (and quickly!) to mitigate the potential Integrity impact of the launch," he wrote, adding that the company hoped that cracking down on engagement bait tactics would help somewhat. Finding a way to measure how much damage MSI caused to the platform's integrity efforts would go on the team's "wish list."

"When an engineer tells you, 'We'll get to that,' you have to understand they're lying," noted the former director who worked on the MSI change.

The company, of course, did not talk about these concerns in public. As proof of the platform's altruistic motives, Zuckerberg said, commenting on the company's fourth-quarter results in 2017, that Facebook would be sticking with MSI even though it was costing the platform 50 million hours of usage per day.

The company's efforts to spin the adoption of MSI as a public good weren't just for external consumption. Whenever employees at a town hall questioned whether Facebook might be prioritizing the welfare of its platform over the welfare of its users, executives would point to Meaningful Social Interactions. How could anyone accuse the company of acting irresponsibly when Zuckerberg had sacrificed so much usage on the altar of bringing people closer together?

The argument largely worked—except among Integrity staffers and the circle of engineers who knew that MSI had been slapped together to fix the problem of *engagement,* a need considered far more serious than boosting time spent.

Facebook still maintained that MSI was indeed a good thing for the world and not rushed, an argument they said was bolstered by undisclosed survey work. But later research found that, as a way to bring people closer together, MSI was useless. Users reported themselves less satisfied with News Feed and less inclined to say that Facebook showed them posts from people important to them. But "business metrics"—how much they boosted usage and commenting—were strong, and the reworked News Feed produced an estimated 0.17 percent increase in daily users.

It would take a year before outsiders began to realize that there was more to Meaningful Social Interactions than met the eye. An analysis by NewsWhip would discover that the adoption of MSI turned the rarely used "angry" emoji into the bellwether of political content's success. Fox News, Breitbart, and the page of *Daily Wire* editor Ben Shapiro led in "angries," the analytics firm wrote, with the liberal *Daily Kos* and Senator Bernie Sanders's Facebook page also gathering them aplenty. Other topics that seemed to out-perform after the shift were abducted children and abortion. *Slate* noted that the most-shared post on the platform in early 2019 was a 119-word news brief with the less-than-informative title "Suspected Human Trafficker, Child Predator May Be in Our Area."

Facebook's after-the-fact research on the effects of MSI only con-firmed these findings. The stories that got clicks were those that trig-gered "negative user sentiment," the company found.

This was not all MSI's fault, of course. Journalism has always thrived on conflict—"if it bleeds it leads," the old journalistic saying went. There was no reason to expect social media to be any different (a pre-MSI report by NewsWhip noted that stories about animal abuse did great on resharing). But the incentives on the platform had dramatically changed, and not just for news publishers.

The effect was more than simply provoking arguments among friends and relatives. As a Civic Integrity researcher would later report back to colleagues, Facebook's adoption of MSI appeared to have gone so far as to alter European politics. "Engagement on positive and policy posts has been severely reduced, leaving par-ties increasingly reliant on inflammatory posts and direct attacks on their competitors," a Facebook social scientist wrote after interview-ing political strategists about how they used the platform. In Poland, the parties described online political discourse as "a social-civil war." One party's social media management team estimated that they had shifted the proportion of their posts from 50/50 positive/negative to 80 percent negative and 20 percent positive, *explicitly as a function of the change to the algorithm.* Major parties blamed social media for deepening political polarization, describing the situation as "unsustainable."

The same was true of parties in Spain. "They have learnt that harsh attacks on their opponents net the highest engagement," the researcher wrote. "From their perspective, they are trapped in an

inescapable cycle of negative campaigning by the incentive structures of the platform."

If Facebook was making politics more combative, not everyone
was upset about it. Extremist parties proudly told the researcher that
they were running "provocation strategies" in which they would
"create conflictual engagement on divisive issues, such as immigra-
tion and nationalism."

To compete, moderate parties weren't just talking more confron-
tationally. They were adopting more extreme policy positions, too.
It was a matter of survival. "While they acknowledge they are contributing to polarization, they feel like they have little choice and are
asking for help," the researcher wrote.

Facebook could ease the pressure to produce bombastic content
by undoing the ranking changes, the moderate parties said—or at
the very least reverting politics-focused pages to the old system.

To the researcher this sounded both sensible and urgent. He lik-
ened MSI to serving users junk food: "its short-term value is not
worth the long-term cost. We should therefore reduce the incentive
for publishers to elicit outrage from their audiences, as it's the right
thing to do for both our mission and our long-term growth." Commenting on those conclusions several years later, Facebook said it
didn't consider the researcher's work convincing.

Later research conducted in Asia would reach a similar conclusion about MSI's caustic effects, but neither the original nor
follow-on work triggered the immediate reckoning the researcher
had called for. Only much later would senior managers such as Guy
Rosen, now Facebook's chief of security, internally acknowledge
that, while distributing content to maximize MSI was "cool for cat
videos," it was a poor proxy for value on more serious topics.

There had been good intent behind some of the early thinking
on MSI, even an honest desire to respond to external criticism that
time on Facebook wasn't time well spent. But the actual metric that
the company had built conflated the good of its users with a business
imperative. The well-being that Facebook promoted was its own.

6

If Jeff Allen had an intuitive understanding of the trolls and opportunists who had degraded Facebook's News Feed, it was because he had once been such a mercenary himself.

Allen grew up in Olathe, Kansas, an upscale suburb of Kansas City, the son of a computer engineer ensconced in the Midwest's "Silicon Prairie" tech scene. He got his first computer at the age of seven. Allen liked computers, but he preferred playing video games. His parents weren't as eager to subsidize that activity, however, limiting his options to a couple dozen Nintendo cartridges he bought with money he made as a checkout boy at the Price Chopper. He earned $5.15 an hour, but games cost around $60.

Technology solved that math problem. In the mid-1990s, a hacker created the first video game emulator, software that allowed a personal computer to mimic the processor of a video game console. Suddenly, bootlegged copies of games began flooding file-sharing websites, obliterating the natural limits of a tech-savvy teenage boy's ability to possess them. *Oh my god, I now need every single Super Nintendo game,* Allen thought. In time, his collection grew to more than nine hundred Nintendo and seven hundred Super Nintendo titles. He uploaded them to a website, Emunation.com, so his friends could play the games, too.

If it weren't for a Silicon Valley innovation, Allen's high school career in copyright infringement might have ended there. But internet ads were about to take over the world. The first one had run in 1994, and standards to run ads were low. At the urging of a friend he met online, Allen applied to DoubleClick, the nascent digital ad giant later bought by Google. Would the company allow his pirated game site to run ads?

DoubleClick was fine with that. All Allen needed to do was paste a snippet of code into his website's HTML and he would get a small cut of the proceeds from whatever ads DoubleClick placed on it. After a few weeks, the company sent Allen, then fifteen, a check—what he referred to as "my first internet money." It was as much as he had made at Price Chopper in three months. *Fuck sacking groceries, I'm doing the internet,* Allen thought.

Allen was hardly the only video game collector to have this thought. Emunation.com was up against dozens of other bootlegged-game sites, all of them competing for attention from a handful of search engines like WebCrawler, AltaVista, and Yahoo. What set Allen apart was that he quickly figured out how to cheat.

"You could manipulate the search engine results just by pasting invisible white text with a particular keyword over and over again at the end of an HTML string," he said. "If you did that and then waited two days you'd be the number one search for that term on WebCrawler," he said. "The whole thing seemed like a game."

The more he fooled the search engines, the more money he made, and soon thousands of dollars were flowing in every month. Between 1997 and 1999, Allen was running what he believes to be the most successful video game emulation website in the world.

The party ended when a letter from Nintendo's American legal counsel landed in his parents' mailbox shortly before Allen's eighteenth birthday. Piracy at the scale Allen was facilitating was ample grounds for a felony charge, the attorney noted. "I was like, 'Still a minor!'" Allen said. "If they'd gotten me a month later, I would have been scared." Allen went straight, shutting the site down and heading to college.

During the years it took Allen to earn a bachelor's degree and then a doctorate in physics, a startup called Google obliterated all its search engine competitors because it figured out how to make the cheating that Allen had excelled at harder. Rather than trusting websites to represent their contents accurately, Google evaluated

whether other websites appeared to trust them, using the intercon-
nections of the open web to distinguish between reputable sources
of information and cleverly optimized trash.

This innovation would have blown high school Jeff Allen out of
the water, but manipulating Google's results wasn't impossible. And
Allen, having concluded that the road to a tenured physics job was
long and bleak, was again on the lookout for "internet money." He
was soon hired by About.com, a web behemoth that created content
about anything that would appear prominently in search results, to
figure out which Google queries it should target and how to structure its site for optimal search result placement.

There was big money to be made playing these games, but Google
clearly wasn't happy about it. Every few months, the search engine
would change its ranking algorithm in a way that tanked About
.com's traffic. The more Allen studied Google's tactics, the more
impressed he became. Google began by dinging his employer for
obvious indications of quality problems, like misspellings, gram-
matical errors, and plagiarism. But the effort to computationally
answer the question "Is this crap?" had grown more sophisticated,
with Google learning to penalize sites that featured redundant content, shallow summaries, and articles by authors whose work was
referenced nowhere else on the internet.

When Google employees would meet Allen's colleagues, they
would explain matter-of-factly that their rules were intended to
undermine the success of the games of the sort that About.com was
playing. Just as Allen was looking for the soft spots in the search
engine's defenses, Google staffers were looking for weaknesses in
About.com's manipulation attempts, and they found them. About
.com responded by devising new ways to game Google, but it was a
losing battle.

With Google slowly cutting off the air supply to the site's chum,
Allen and his bosses had to consider the unthinkable: producing
better content. Not only did About.com start investing more in the
articles it produced; the site started culling the mountain of old
trash content that was dragging its placement in Google results

down. Instead of gaming Google's rules, Allen was now tasked with complying with them. He began writing algorithms that identified and deleted the site's most pointless and sloppy content.

Over the course of two years, the number of published articles on About.com fell from more than 4 million to 700,000. The remaining content wasn't going to win any literary prizes, but it was unquestionably more useful. In response, Google search traffic came back.

About.com's Google-mandated shift toward quality helped extend the life of its business, but the lesson to be learned wasn't that good content always outperformed. It was just not to mess with Google.

On other platforms, About.com's quest to reap page views for junk content had continued unabated. One notable success came on Facebook, when About.com realized it could profitably buy ads promoting its own ad-laden content. One in particular, a celebrity slideshow titled "Should I Get a Pixie Haircut," proved to be a gold mine when set on repeat—the elderly target demographic wouldn't realize they had hit the end and would just keep clicking.

But Allen was no longer at About.com to share the glory when a former colleague told him that the site had discovered how to turn an infinite loop of Dame Judi Dench and Charlize Theron glamour shots into truckloads of money. By mid-2016, he'd taken a data science job at Facebook.

Allen's first role at Facebook was improving the platform's local search results, an effort intended to help compete with Google's business search function and Yelp. But, although people regularly connected with businesses they knew on Facebook, they rarely went looking for new ones. "When you typed in an actual local search query that actual humans type into actual search engines, our local search engine just utterly fell on its face," Allen said.

Allen was pretty sure this was a problem, but his bosses questioned that assessment. If users wanted to find the types of local businesses that Allen said they did, why weren't those search queries showing up in the company's user data? To Allen, there was a clear catch-22 at play. Facebook's local search results sucked, and they wouldn't improve until the company had the behavioral data to improve them. But users weren't generating that data, because Facebook's local search results sucked.

Solving the problem would have been hard in any circumstance, but it was impossible given the skepticism from higher-ups. Allen had run into a blind spot of Facebook's—the company's supreme confidence that mountains of behavioral data revealed what users wanted from its platform, rather than just their response to the limited options they were given.

By 2018, Allen had gotten bored and transferred to Facebook's Page Ranking team. His focus wasn't supposed to be integrity work per se, but Allen immediately saw a problem: Facebook's most successful publishers of political content were foreign content farms posting absolute trash, stuff that made About.com's old SEO chum look like it belonged in the *New Yorker*.

Allen wasn't the first staffer to notice the quality problem. The pages were an outgrowth of the fake news publishers that Facebook had battled in the wake of the 2016 election. While fact-checks and other crackdown efforts had made it far harder for outright hoaxes to go viral, the publishers had regrouped. Some of the same entities that BuzzFeed had written about in 2016—teenagers from a small Macedonian mountain town called Veles—were back in the game. How had Facebook's news distribution system been manipulated by kids in a country with a per capita GDP of $5,800?

Allen had a ready answer to that question. He didn't need to board a flight to know *exactly* who it was sitting in front of those computers: high-school-aged Slavic Jeff Allens.

Everything lined up. Two decades earlier, Yahoo and its ilk had failed to grasp how scrubs like Allen would trial-and-error their way into a rough understanding of their algorithms, then use that knowledge to break them for profit. The search engines had never recovered.

Facebook had made the same mistake, underestimating the threat of nihilistic yet tech-savvy teenagers far from Silicon Valley. Allen almost admired them.

"When I saw the troll farms, I didn't blame them," Allen said. "I was like, 'Fuck yeah, bros—make that money! If I was seventeen I'd be with you, 100 percent.'"

But Allen was no longer seventeen. He had a more fully developed moral compass. And, with all due respect to copyright law, the harm of commercializing social and political disinformation at a mass scale was a lot greater than pissing off Nintendo's lawyers.

Allen's experience told him things would get worse. It had taken time for people like him to figure out how to game search engines— but once the secrets were discovered, that knowledge spread like a virus. You didn't need to be an elite cyber operative or social media prodigy. Once the codes got cracked, anyone could copy them.

By the time Allen transferred to Facebook's Pages team, the News Feed ranking staffers investigating Russian propaganda in the wake of the election had identified a frequent tactic in their growth: ads.

Facebook's advertising system let businesses choose what they wanted to accomplish. Coca-Cola might want a huge audience to see its brand, and nothing more. A band might want users to visit its website and sign up for emails about upcoming show dates. A mattress company might want to pay Facebook only for the attention of users who eventually made a purchase. To help businesses meet these varying goals, Facebook let advertisers choose whether they paid for impressions, for clicks, or for verified sales.

There were also Page Like ads—the kind that Russians had used to good effect. Their $100,000 investment, purchased by the so-called Internet Research Agency, had allowed them to grow their pages at a rate that organic growth could never have rivaled. "The IRA used Facebook the way it's designed to be used," recalled one Facebook staffer assigned to post-election forensics. Buying the ads had solved a cold start problem, giving the Russians a seed audience they could nurture to gargantuan size.

Page Like ads were hardly a cornerstone of Facebook's business, accounting for a low single-digit percentage of the company's revenue. But they had been a foundational tool for launching not just the Russian election interference but countless other nickel-and-dime manipulation schemes.

So why not, Allen wondered, stop selling gasoline to arsonists?

His team had tried a year earlier, proposing that the ads either be restricted or discontinued entirely. Predictably, there were two major constituencies opposed to the idea—the team paid to sell ads and the team paid to make pages grow. The idea never received serious consideration and was off the table by the time Allen joined the team in 2018.

Allen had reached what would turn out to be a rite of passage for integrity-focused staffers: realizing that business pressures had taken the simplest and most direct solution to a problem off the table. Some of his more earnest colleagues struggled to overcome that demoralizing hurdle, but Allen's entire career was premised on finding ways around designed constraints. Blocked from attacking unscrupulous pages' early-stage growth tactics, he began studying other ways in which they were gaming Facebook.

Allen needed to learn from the masters—publishers who knew nothing about American audiences, spent nothing on content, and yet succeeded on Facebook. In other words, he needed to learn from the Macedonians.

When reviewing troll farm pages, he noticed something—their posts usually went viral. This was odd. Competition for space in users' News Feeds meant that most pages couldn't reliably get their posts in front of even those people who deliberately chose to follow them. But with the help of reshares and the News Feed algorithms, the Macedonian troll farms were routinely reaching huge audiences. If having a post go viral was hitting the attention jackpot, then the Macedonians were winning every time they put a buck into Facebook's slot machine.

The reason the Macedonians' content was so good was that it wasn't theirs. Virtually every post was either aggregated or stolen from somewhere else on the internet. Usually such material came from Reddit or Twitter, but the Macedonians were just ripping off content from other Facebook pages, too, and reposting it to their far larger audiences. This worked because, on Facebook, originality wasn't an asset; it was a liability. Even for talented content creators, most posts turned out to be duds. But things that had already gone viral nearly always would do so again.

The Macedonians were gaming Facebook with cut-and-pasted content in much the same way that Allen had gamed WebCrawler by coding invisible keywords into his website's HTML. And, like him, they were getting paid via ad money, funneling their audience to content farms or collecting payouts from a Facebook revenue-sharing program for large on-platform publishers. But the Macedonians had a third, more ominous way of earning their internet money. They could sell their big pages to buyers with unknown intentions.

Allen began a note about the problem from the summer of 2018 with a reminder. "The mission of Facebook is to empower people to build community. This is a good mission," he wrote, before arguing that the behavior he was describing exploited attempts to do that. As an example, Allen compared a real community—a group known as the National Congress of American Indians. The group had clear leaders, produced original programming, and held offline events for Native Americans. But, despite NCAI's earnest efforts, it had far fewer fans than a page titled "Native American Proub" [sic] that was run out of Vietnam. The page's unknown administrators were using recycled content to promote a website that sold T-shirts.

"They are exploiting the Native American Community," Allen wrote, arguing that, even if users liked the content, they would never choose to follow a Native American pride page that was secretly run out of Vietnam. As proof, he included an appendix of reactions from users who had wised up. "If you'd like to read 300 reviews from real users who are very upset about pages that exploit the Native American community, here is a collection of 1 star reviews on Native American 'Community' and 'Media' pages," he concluded.

This wasn't a niche problem. It was increasingly the default state of pages in *every* community. Six of the top ten Black-themed pages—including the number one page, "My Baby Daddy Ain't Shit"—were troll farms. The top fourteen English-language Christian- and Muslim-themed pages were illegitimate. A cluster of troll farms peddling evangelical content had a combined audience twenty times larger than the biggest authentic page.

"This is not normal. This is not healthy. We have empowered inauthentic actors to accumulate huge followings for largely unknown purposes," Allen wrote in a later note. "Mostly, they seem to want to skim a quick buck off of their audience. But there are signs they have been in contact with the IRA."

So how bad was the problem? A sampling of Facebook publishers with significant audiences found that a full 40 percent relied on content that was either stolen, aggregated, or "spun"—meaning altered in a trivial fashion. The same thing was true of Facebook video content. One of Allen's colleagues found that 60 percent of video views went to aggregators.

The tactics were so well-known that, on YouTube, people were putting together instructional how-to videos explaining how to become a top Facebook publisher in a matter of weeks. "This is where I'm snagging videos from YouTube and I'll re-upload them to Facebook," said one guy in a video Allen documented, noting that it wasn't strictly necessary to do the work yourself. "You can pay 20 dollars on Fiverr for a compilation—'Hey, just find me funny videos on dogs, and chain them together into a compilation video.'"

Holy shit, Allen thought. Facebook was losing in the later innings of a game it didn't even understand it was playing. He branded the set of winning tactics "manufactured virality."

"What's the easiest (lowest effort) way to make a big Facebook Page?" Allen wrote in an internal slide presentation. "Step 1: Find an existing, engaged community on [Facebook]. Step 2: Scrape/ Aggregate content popular in that community. Step 3: Repost the most popular content on your Page."

Thankfully, Allen noted, these tactics were easy to detect. The pages of trash publishers consistently lacked named authors or even contact information, and their content went viral too predictably, like a roulette player winning on every spin of the wheel. On top of it all, Facebook already had systems to fingerprint and flag individual pieces of content—it used them for such things as automatically blocking images of child sexual abuse the moment they were uploaded.

If Facebook couldn't shut garbage page proprietors down, he added, it could make their work a lot harder. "We should remove content that exploits communities from our platform, or remove all motivation for putting it on our platform because it is antithetical to our mission," Allen wrote to colleagues in the summer of 2018.

Allen's research kicked off a discussion. That a top page for American Vietnam veterans was being run from overseas—from Vietnam, no less—was just flat-out embarrassing. And unlike killing off Page Like ads, which had been a nonstarter for the way it alienated certain internal constituencies, if Allen and his colleagues could work up ways to systematically suppress trash content farms—material that was hardly exalted by any Facebook team—getting leadership to approve them might be a real possibility.

This was where Allen ran up against that key Facebook tenet, "Assume Good Intent." The principle had been applied to colleagues, but it was meant to be just as applicable to Facebook's billions of users. In addition to being a nice thought, it was generally correct. The overwhelming majority of people who use Facebook do so in the name of connection, entertainment, and distraction, and not to deceive or defraud. But, as Allen knew from experience, the motto was hardly a comprehensive guide to living, especially when money was involved.

AltaVista, WebCrawler, and the other search engines that Allen had gamed had failed because they hadn't been prepared for websites to cheat their way to the top. Google had succeeded because its founders had recognized that the intentions of a young Jeff Allen and entities like About.com weren't to be trusted.

"It always bothered me when Zuckerberg said that the abuse of Facebook couldn't be predicted, that he couldn't see the future from a Harvard dorm room," Allen said. "Well, the future looked real obvious from a Palo Alto garage."

The road to Facebook's current hell was paved with Assuming Good Intent. Allen could hardly rescind a line in the company's employee handbook, but he could at least try to teach his colleagues from personal experience.

He put together a slide deck titled "Confessions of an SEO Scumbag," using the industry acronym for "search engine optimization." You didn't have to be a Russian operative to ruin Facebook, Allen noted. You just needed to be a little nihilistic and have a taste for what sixteen-year-old Jeff Allen had referred to as "internet money."

Targeting specific pages of parasites like himself might provide temporary respite, but the only plausible path to defeat them was requiring substance. There was no way around it. Facebook was going to have to address quality.

"For the past nine months, I've been studying and researching existing web platforms and the history of media over the last 150 years," Allen wrote in his presentation. "We own the platform. We play games on our terms."

With the help of another data scientist, Allen documented the inherent traits of crap publishers. They aggregated content. They went viral too consistently. They frequently posted engagement bait. And they relied on reshares from random users, rather than cultivating a dedicated long-term audience.

None of these traits warranted severe punishment by itself. But together they added up to something damning. A 2019 screening for these features found 33,000 entities—a scant 0.175 percent of all pages—that were receiving a full 25 percent of all Facebook page views. Virtually none of them were "managed," meaning controlled by entities that Facebook's Partnerships team considered credible media professionals, and they accounted for just 0.14 percent of Facebook revenue. If Facebook wanted to encourage good publishers—entities with reputable owners who invested in producing their own content—it needed to discriminate against those that were gaming the platform to manufacture virality. Allen was certain the approach would work: it was pretty similar to the Google tactics that had forced About.com to clean up.

It was a compelling argument, but it was going to be a tough sell: the last thing Facebook, and Zuckerberg in particular, wanted was to adjudicate quality.

Ten years and tens of thousands of hostile articles ago, Facebook had been a source of hope to the journalism business. Strategists in newsrooms across America had presented graphs of rising digital ad dollars offsetting deteriorating revenues from print advertising and subscriptions, promising to hasten the day the lines converged. Although making up lost subscription dollars from digital ad pennies was a stretch for most publications, traffic from Facebook had a jackpot-like quality. When a story landed on the platform, it could easily earn ten or a hundred times the page views it would have otherwise.

It was just as digital outlets truly started digging for social media traffic gold that Brandon Silverman happened to get into the business of selling shovels.

Silverman was a progressive with an enthusiasm for digital organizing. In 2011, he had cofounded a company called CrowdTangle with the goal of helping nonprofits and lefty political groups design and manage online activism campaigns. The product flopped with its target market, but it proved unusually good at tracking what was happening on platforms like Facebook and Twitter. If you wanted to know when a topic began drawing unusual traffic or spot a post at the beginning of a viral growth curve, CrowdTangle could do that. The capacity to act as a virality sniffer was an immensely valuable feature at the dawn of the attention economy, when marketing agencies were scrambling to help their clients generate social media buzz and track their mentions.

Silverman had stumbled into a gold mine, and there was only one thing standing between him and a lucrative career in corporate brand monitoring: he considered it to be deeply lame. "I came home one night and told my wife, 'If this thing is going to be a social analytics tool, you have to shoot me in the head,'" he said.

A less profitable but more personally appealing option for Silverman was working with news publishers. Even the stodgiest news outlets were beginning to get curious about what was trending on social media, and a new crop of digital native publishers like BuzzFeed and Upworthy were racing ahead.

The two business models—consultant both to brands and to news organizations—coexisted within CrowdTangle until 2014, when Silverman was traveling from his DC home to New York to give a demonstration of the tool to Pfizer's marketing team. While on Amtrak, a colleague informed him that NPR also wanted a demo at their DC headquarters, and they were free that same afternoon. "On the one hand, there's potentially millions of dollars from Pfizer," he said. "On the other hand, it was NPR." He got off the train at the next stop and headed back to DC.

CrowdTangle's facility at discovering and tracking viral social media content made it an all too obvious acquisition target. When Silverman showed the tool to the *New York Times,* the first question from the paper's digital team was how long until Facebook bought his company.

Facebook was already an enthusiastic client, as were YouTube, Reddit, and Twitter. One might think these companies would be interested in the tool to learn what their competitors were up to, but it turns out that they were often just as excited to use CrowdTangle to track how various publishers were performing on their own platforms. This struck Silverman as a little ridiculous. Major social media companies were paying CrowdTangle to access the same data that they gave his company for free.

What Silverman was selling wasn't data; it was comprehensibility. Each company had partnership teams—people responsible for liaising with media outlets, celebrities, and sports leagues—who needed to be able to answer questions about what content was outperforming, when and how much to post, and why particular items had gone viral. But the companies they worked for had never built simple internal tools to do those tasks.

"It was the era of big data vibes," Silverman said. "The platforms had built these incomprehensible dashboards with volume charts and needles, but the people we were working with just needed content stacked with clear numbers."

There had been a few flirtations between CrowdTangle and Facebook, but nothing serious until Zuckerberg and Sandberg took a trip in 2016 to Sun Valley, Idaho, for Allen & Co's annual media, entertainment, and technology conference, where media mogul Rupert Murdoch was delivering a talk.

"For the whole session, Rupert just railed on Facebook and why they were a cancer on the news industry," said Silverman, who wasn't there but heard plenty about it from other attendees. "And I think there were a lot of nods and assents."

Privately, Murdoch and Robert Thomson, CEO of Murdoch's News Corp, hosted Zuckerberg at the former's Sun Valley villa to deliver a much more direct warning: if Facebook didn't start working more collaboratively with the news industry, News Corp would become an avowed enemy, pushing against the company on multiple continents.

The threat alarmed Zuckerberg. Facebook hadn't meant to piss off newspapers, but it hadn't been doing much to make friends, either. During a period of paranoia about competition with Twitter in the early teens, Facebook had flooded its platform with news content—a boon for publishers that had been struggling to make digital journalism work. Publishers had invested heavily in the platform—only to realize that, once Facebook's competitive concerns eased, the traffic became fickle. One month, a publication might celebrate getting 30 million views from Facebook traffic, only to see its numbers cut in half the next. Facebook hadn't really considered the threat that pissing off news organizations could potentially pose to its reputation.

To try to mend fences, Facebook hired Campbell Brown, a prominent television journalist, to run a new team that would conduct media diplomacy and build tools that might make Facebook a more reliable partner. There was another part of the plan: Facebook would buy CrowdTangle.

When Facebook approached the company in the summer of 2016, it had a firm price and rationale for the acquisition. Crowd-Tangle would try to help old media institutions thrive on Facebook—or at the very least understand the force disrupting their industry. Silverman and his cofounder accepted the offer.

By the time of CrowdTangle's acquisition, Twitter was no longer considered a significant competitive threat, and Facebook was slashing how much attention it gave to news in one algorithm change after another. Compounding the frustration of major news outlets was that they were getting beaten so heavily by upstart publishers.

"[*Washington Post* executive editor] Marty Baron would go sit down with Zuck," Silverman said, and ask him how it was that "you're sending more traffic to Breitbart versus *Washington Post*," he said. "And Mark would say, 'We show them Breitbart versus *Washington Post* and they choose Breitbart. What are we supposed to do?'"

After it was bought, CrowdTangle was no longer a company but a product, available to media companies at no cost. However much publishers were angry with Facebook, they loved Silverman's product. The only mandate Facebook gave him was for his team to keep building things that made publishers happy. Savvy reporters looking for viral story fodder loved it, too. CrowdTangle could surface, for instance, an up-and-coming post about a dog that saved its owner's life, material that was guaranteed to do huge numbers on social media because it was already heading in that direction.

CrowdTangle invited its formerly paying media customers to a party in New York to celebrate the deal. One of the media executives there asked Silverman whether Facebook would be using Crowd-Tangle internally as an investigative tool, a question that struck Silverman as absurd. Yes, it had offered social media platforms an early window into their own usage. But Facebook's staff now out-numbered his own by several thousand to one. "I was like, 'That's ridiculous—I'm sure whatever they have is infinitely more powerful than what we have!'"

It took Silverman more than a year to reconsider that answer.

Before CrowdTangle could become a boon to company insiders, it had to do what it was purchased for: spread to as many news outlets as possible.

Even as other members of the Partnerships team became peren-nial bearers of bad news, informing publishers of algorithm changes that rarely were in their favor, CrowdTangle provided a respite. If Facebook wasn't a platform on which old-guard media outlets were going to succeed, the product at least showed that the company wanted to help them try.

After a year or so, however, CrowdTangle was approaching saturation—most of the media outlets that could be persuaded to use the product already did, and there were only so many trainings and check-ins that newsroom analytics staff and social media managers could use. Silverman, therefore, began broadening his focus, promoting the tool to entities that wanted to *understand* viral content on Facebook, not just produce more of it.

One such audience was reporters. It turned out that the tool was great at tracking the spread of false claims, conspiracy groups, and viral rumors. The power to easily quantify such problems could make for discoveries that embarrassed Facebook, though. Craig Silverman, the BuzzFeed reporter (no relation to Brandon Silverman), had used the tool to help expose the extent of Facebook's news problem. But the company didn't waver. CrowdTangle benefited from both a genuine internal recognition that scrutiny could make its products better and the fulsome support of Campbell Brown, who was trying to navigate her staff through hard conversations with media outlets.

A second audience was Facebook's own researchers and Integrity staff. The company had, of course, vastly more data than what was built into CrowdTangle, but Silverman found that it often wasn't put to use. At times even executives struggled to grasp the range of what was happening on Facebook, since they used the product sparingly or in ways that didn't reflect the range of the ecosystem of videos, pages, and groups that now existed on the platform. CrowdTangle could be used to fill in some of those blind spots and collect quick and dirty intelligence.

The tool could even be used to detect emerging threats against company executives made on its platforms—an item of concern after an April 2018 shooting at YouTube left three people injured. It could produce automated briefings for regional content moderation centers so staff would have context for what was happening on the platform each day. And when Facebook did want to look into a specific issue—say, groups devoted to trading exotic animals, a violation of platform policies—CrowdTangle offered a quick screening tool.

It was only as CrowdTangle started building tools to do this that the team realized just how little Facebook knew about its own platform. When Media Matters, a liberal media watchdog, published a report showing that MSI had been a boon for Breitbart, Facebook executives were genuinely surprised, sending around the article asking if it was true. As any CrowdTangle user would have known, it was.

Silverman thought the blindness unfortunate, because it prevented the company from recognizing the extent of its quality problem. It was the same point that Jeff Allen and a number of other Facebook employees had been hammering on. As it turned out, the person to drive it home wouldn't come from inside the company. It would be Jonah Peretti, the CEO of BuzzFeed.

BuzzFeed had pioneered the viral publishing model. While "listicles" earned the publication a reputation for silly fluff in its early days, Peretti's staff operated at a level of social media sophistication far above most media outlets, stockpiling content ahead of snowstorms and using CrowdTangle to find quick-hit stories that drew giant audiences.

In the fall of 2018, Peretti emailed Cox with a grievance: Facebook's Meaningful Social Interactions ranking change was pressuring his staff to produce scuzzier content. BuzzFeed could roll with the punches, Peretti wrote, but nobody on his staff would be happy about it. Distinguishing himself from publishers who just whined about lost traffic, Peretti cited one of his platform's recent successes: a compilation of tweets titled "21 Things That Almost All White People Are Guilty of Saying." The list—which included "whoopsie daisy," "get these chips away from me," and "guilty as charged"—had performed fantastically on Facebook. What bothered Peretti was the apparent reason why. Thousands of users were brawling in the comments section over whether the item itself was racist.

"When we create meaningful content, it doesn't get rewarded," Peretti told Cox. Instead, Facebook was promoting "fad/junky science," "extremely disturbing news," "gross images," and content that exploited racial divisions, according to a summary of Peretti's email that circulated among Integrity staffers. Nobody at BuzzFeed liked producing that junk, Peretti wrote, but that was what Facebook was demanding. (In an illustration of BuzzFeed's willingness to play the game, a few months later it ran another compilation titled "33 Things That Almost All White People Are Guilty of Doing.")

Cox was intrigued. He made an offer: join Facebook and help the company fix the problem. Peretti told Cox he would do it, but only if Facebook bought BuzzFeed. Cox floated the idea to other Facebook executives, but it was dead in the water.

"We can't buy BuzzFeed," he told Peretti.

"Then I can't come work with you," Peretti responded.

The incident nonetheless helped spur Cox to start pushing the issue of declining quality himself. He asked Tom Alison, vice president of Engineering, to try to come up with a functional definition of content quality and an idea of what Facebook wanted to promote on its platform. Alison turned to Silverman at CrowdTangle for help. Suddenly, Silverman had an opening to show executives what they were missing.

That was how, in late 2019, Silverman found himself in a conference room walking the product heads for News Feed, Video, and Groups through the day's top content, post by post, content that any CrowdTangle user could look at but that Facebook's executives didn't tend to see. There was some misinformation on the list Silverman shared—that day an account was going viral impersonating Britney Spears—though it didn't really stand out. The unifying trait was that all of it was awful. Some posts were removal-worthy, others merely vulgar, aggregated, or spammy. None of it was the kind of "meaningful" content that Facebook sought to reward.

Executives shifted in their seats. The problem wasn't just News Feed. Groups, pages, and other products had all played a role.

The room was already plenty uncomfortable before someone—Silverman can't recall if it was him—asked if everyone felt proud of the list.

"There was just the most pregnant, heavy pause of any room I've ever been in," Silverman said. "Then everyone moved on to another question."

—

If Silverman was trying to make Facebook's leaders take action by forcing them to pay attention to its top content, members of Facebook's Partnerships team were making a similar argument by pointing out the sort of material that had dropped off the list. The platform hadn't been built with celebrities, musicians, and other public figures in mind, but, as it had grown, it had become hard for them to ignore. For entertainers looking to connect directly with fans, it became especially valuable, allowing them to inject announcements straight into their audience's feeds.

Those conditions were, however, too good to last. As users' News Feeds became dominated by reshares, group posts, and videos, the "organic reach" of celebrity pages began tanking. "My artists built up a fan base and now they can't reach them unless they buy ads," groused Travis Laurendine, a New Orleans–based music promoter and technologist, in a 2019 interview. A page with 10,000 followers would be lucky to reach more than a tiny percent of them.

Explaining why a celebrity's Facebook reach was dropping even as they gained followers was hell for Partnerships, the team tasked with providing VIP service to notable users and selling them on the value of maintaining an active presence on Facebook. The job boiled down to convincing famous people, or their social media handlers, that if they followed a set of company-approved best practices, they would reach their audience. The problem was that those practices, such as regularly posting original content and avoiding engagement bait, didn't actually work. Actresses who were the center of attention on the Oscars' red carpet would have their posts beaten out by a compilation video of dirt bike crashes stolen from YouTube.

"By 2019, it had gotten hard to tell a public figure why they should do anything on Facebook," recalled a former employee who worked with brand-name performers. "Musicians wanted to sell tickets, authors wanted to sell books. And we weren't helping them with that."

The situation was so dire that the Partnerships team approached Zuckerberg with a "hierarchy of needs" for celebrities on the platform, according to the former celebrity liaison. At the base of the pyramid was the need to reach one's own fans—without that, no amount of white-glove service or special dashboards was going to help. Their message didn't really get through. The feeling was that "the algorithm would do its job to feed people the right thing at the right time," the person said. For Facebook's partners, successful content was something you had to invest in. For Facebook, it was just something that happened.

Brian Boland, the former Partnerships VP, remembered one instance in particular that illustrated the type of thinking that was prevalent on the product side of Facebook. When the company introduced the ability for users to post video, Boland recalled, the executives in charge of the project excitedly told him the new format was going bonkers. When he asked what seemed like an obvious follow-up question—what *types* of video were succeeding?—they didn't know.

Over time, celebrities and influencers began drifting off the platform, generally to sister company Instagram. "I don't think people ever connected the dots," Boland said.

7

With Facebook still enmeshed in scandals, even as the 2018 midterm elections were swinging into gear, the company knew it needed to publicly prove that it was capable of maintaining order on its platform. One component of that effort was demonstrating that the company could find and eliminate foreign troll farms. The other was showing off the work of its Civic Integrity team.

Civic was, in some ways, an odd choice for flaunting Facebook's election defense prowess. For the first two years of its existence, the team had overwhelmingly focused on making Facebook better for democracy—not preventing the platform from undermining it. They built things like "constituent badges," which let members of Congress see if a user engaging with them was a resident of their district. (Whatever the utility to legislators, the badge changed user behavior. Comments left by badged users tended to be more substantive, and women, who interacted with their representatives significantly less often than men, suddenly began commenting more.)

Samidh Chakrabarti, who oversaw the team's product development, had joined Facebook in 2015 after working at Google, where he ran the company's efforts to provide users with voting information, registration opportunities, and polling place reminders.

Chakrabarti was enthusiastic about the future of democracy and the internet. After launching the constituent badge, he had told reporters and staffers, "I've always been fascinated by how the internet helps citizens have a voice like never before." He told colleagues he joined Facebook because of the company's commitment to both civic engagement work and to studying itself. Where Google was reluctant to go beyond registering voters and reminding people it was Election Day, Facebook wanted to show that its voting "nudges" had a statistically significant effect on turnout and then find new ways to demonstrate the company's growing centrality in politics.

There had been plenty to do on that front. Facebook's voter registration program could be improved. It was tinkering with possible formats for online town halls and it wanted to build more features for civic organizing. Ahead of the next election, the team had released a rigorously tested "ballot preview" feature that increased voters' knowledge about campaigns and ballot initiatives by 6 percent.

To guide the work, Chakrabarti and his team had put together a list of six principles, an informal "civic oath." It read: "Be Selfless, Be Protective, Be Fair, Be Representative, Be Constructive, Be Conscious." These exhortations might recall construction-paper directives tacked to an elementary school wall, but, even in those optimistic days, they were defined with a bit of a bite.

"Be Selfless" meant that the team was there to "serve people's interests first, not Facebook's interests." "Be Constructive" came with a mandate to build empathy and minimize acrimony in the public sphere. And "Be Conscious" was a call to document how Facebook made the world better *and* worse.

Those who joined Civic tended to be liberal. This was not demographically surprising for the company at large. From Facebook's Menlo Park headquarters, the nearest county that voted for Trump was a two-hour drive, and federal fundraising records show the company's employees favor Democrats seven to one. But the skew was likely even greater among those drawn to working on subjects related to misinformation, foreign interference, and politics. This was unquestionably awkward given the company's responsibility to be nonpartisan. The team was composed of liberals who had a responsibility to be neutral.

At meetings every six months, Chakrabarti would give out awards, nicknamed "Civvies," to staffers whose work had exemplified the team's self-appointed mission. Among the earnest politics nerds drawn to the team, the awards were a moderately big deal, nearly on par with the "Performance Summary Cycle" reviews that determined Facebook employee bonuses. Outside of the team, nobody cared.

As much as Civic's values acknowledged the potential for Facebook to be a force for ill, the team was, in its early life, generally focused on the upside. Helping people use the platform more effectively for political activism required understanding how they were using it already, and in early 2016 a few staffers began studying highly successful activist pages and groups to understand the winning approach.

The results weren't pretty. The most popular American political groups on Facebook weren't American, and they damn sure weren't motivated by a passion for democracy. Civic had found the Macedonians long before foreign troll farms became a topic of public concern.

One of the same staffers did another round of work in April 2016, analyzing the platform's most successful German political groups. She found that a third of them routinely featured content that was racist and conspiracy-minded—and they also tended to be oddly pro-Russian. This was bad, but what was worse was the discovery that these groups were heavily dependent for their growth on Facebook's "Groups You Should Join" and "Discover" suggestions. Not only were both the Macedonian pages and German extremist groups big on the platform; they owed their growth to Facebook's own mechanics.

"Sixty-four percent of all extremist group joins are due to our recommendation tools," the researcher wrote in a note summarizing her findings. "Our recommendation systems grow the problem."

This sort of thing was decidedly not supposed to be Civic's concern. The team existed to promote civic participation, not police it. Still, a longstanding company motto was that "Nothing Is Someone Else's Problem." Chakrabarti and the researcher team took the findings to the company's Protect and Care team, which worked on things like suicide prevention and bullying and was, at that point, the closest thing Facebook had to a team focused on societal problems.

Protect and Care told Civic there was nothing it could do. The accounts creating the content were real people, and Facebook intentionally had no rules mandating truth, balance, or good faith. This wasn't someone else's problem—it was nobody's problem.

Once the 2016 election had passed and all these issues were pushed to the surface, Chakrabarti set up a meeting with Cox. His small team had dusted off its research on the Macedonians and given it an update, finding that, in the seven or eight months since the last analysis, the already-sizable groups had grown massively.

Cox's interest in polarization was well-known, and Chakrabarti walked him through slides showing how a small group of pages had managed to pump out low-budget viral hits that dwarfed the engagement of legitimate digital publishers. Chakrabarti would later tell his team that Cox was shocked by the Macedonians' outperformance. "We've got to work on this," Cox said, according to people Chakrabarti spoke to afterward.

The presentation next went to Sandberg, who seemed to understand that fake news was both a legitimate problem and a serious public relations threat. The News Feed Ranking team was already at work on quantifying the misinformation problem and thinking of ways to fight it, but nobody was really looking at the rest of the platform, the ecosystem of political groups and pages that helped speed such content along its way.

Even if the problem seemed large and urgent, exploring possible defenses against bad-faith viral discourse was going to be new territory for Civic, and the team wanted to start off slow. Cox clearly supported the team's involvement, but studying the platform's defenses against manipulation would still represent moonlighting from Civic's main job, which was building useful features for public discussion online.

A few months after the 2016 election, Chakrabarti made a request of Zuckerberg. To build tools to study political misinformation on Facebook, he wanted two additional engineers on top of the eight he already had working on boosting political participation.

"How many engineers do you have on your team right now?" Zuckerberg asked. Chakrabarti told him. "If you want to do it, you're going to have to come up with the resources yourself," the

CEO said, according to members of Civic. Facebook had more than 20,000 engineers—and Zuckerberg wasn't willing to give the Civic team two of them to study what had happened during the election.

Chakrabarti reassigned two members of his staff instead and begged for assistance from members of the Core Data Science division, the closest thing Facebook had to a tenured-professor track outside of its AI research operation. Civic had no problem recruiting help there. For people who studied human behavior using quantitative methods, the question of how foreign hoax publishers and plagiarists had achieved such success was fascinating.

By the end of 2017, Chakrabarti was overseeing a staff of nearly one hundred. Civic was officially part of Facebook's Media team, which existed to facilitate public figures' use of the platform—an arbitrary placement but one that ultimately fell under Cox's purview, and having Cox's support was key. Facebook's head of Product had championed and directed resources toward not just Civic but any team focused on understanding the platform's vulnerabilities and unintended consequences. Cox seemed aware that he was crucial to such work getting done: shortly before he took a three-month parental leave in 2017, he sent a note to everyone working on integrity, admitting to "a fair amount of anxiety" about spending so much time out of the office.

"We're in tough times," he wrote, thanking the "many of you [who] have been dedicating nights and weekends to integrity efforts over these past few weeks while under fire."

Their work was the most important the company was doing, he wrote, and before he headed out Cox had one final request. "Dig deeper on quality," he wrote. "It's an area I know we can do better."

The model that Facebook should aspire to, Cox wrote, was a manufacturing technique pioneered by Toyota titled "Stop the Line." The premise was that every employee was supposed to halt the entire production line when they saw a defect. The technique produced constant stoppages at first, inconveniencing everyone, but it was the fastest way to identify the root causes of recurring problems.

Integrity staffers should consider themselves part of Facebook's production line, he said, and they should slam on the brakes whenever they saw the company doing slipshod work. Only by improving process and taking the time necessary to identify sources of failure would the company be able to demonstrate its stewardship of democracy and public discourse.

There was no need to point out that this message ran contrary to "Move Fast and Break Things."

The questions Civic was seeking to answer would soon become less academic. Rather than consulting with experts on, for instance, what makes an online town hall a constructive forum, staffers were building screening tools to look for abnormal user behavior ahead of Alabama's special Senate election in December. Among the ranks of new hires, staffers with law enforcement backgrounds became more common—contributing to a sense that the platform's risks were real and the stakes high.

After its initial snub by Zuckerberg, Civic was becoming an increasingly visible part of Facebook's post-election cleanup effort. Chakrabarti became a regular fixture on the company's Hard Questions blog, writing posts like "What Effect Does Social Media Have on Democracy?," published in January 2018. "I wish I could guarantee that the positives are destined to outweigh the negatives, but I can't," he wrote. "That's why we have a moral duty to understand how these technologies are being used and what can be done to make communities like Facebook as representative, civil and trustworthy as possible."

Accompanying the post was a video in which Chakrabarti, a square-jawed man with graying close-cropped hair, spoke earnestly into the camera about why he believed the company could rise to the challenge. "What gives me hope about the future is that the same ingenuity that made Facebook an incredible way to connect with friends and family can also be applied to making it an important and powerful way for people to engage with their communities," he said.

Filmed in a sunny room and set to a softly upbeat piano soundtrack, the video's acknowledgment that Facebook might not be an inherent boon to democracy was hardly radical by 2018. But it was still a big deal for a Facebook official to say it. For all the executives' utopian musings about the platform's capacity to avert wars, quash hatreds, and birth democracies, they had never considered the possibility that it could do harm on the same scale.

While acknowledging the possibility that social media might not be a force for universal good was a step forward for Facebook, discussing the flaws of the existing platform remained difficult even internally, recalled product manager Elise Liu.

"People don't like being told they're wrong, and they especially don't like being told that they're morally wrong," she said. "Every meeting I went to, the most important thing to get in was 'It's not your fault. It happened. How can you be part of the solution? Because you're amazing.'"

Taking on the role of in-house pessimists required some sacrifices. As Civic began shifting toward election defense, the team didn't have much time for the pro-democracy feature-building work it was created to do. Resources were stretched thin, and managers' attention even thinner. Something had to go.

The decision wasn't a close call—studying networks of shill pages trumped enabling more constructive political discourse. Still, dropping that work was painful. At Chakrabarti's request, Harbath's team picked up the constituent badges feature. Almost everything else from its original mission was abandoned.

"All Samidh was able to save was the election reminders and voter registration stuff," Harbath said. "The proactive work went down the toilet, and I think that was really hard for him."

Another big change awaited the team. In a reorganization conducted in early 2018, Civic was uprooted from within Cox's organization and transferred into the company's nascent Central Integrity division, run by Guy Rosen. This was a big shift: where Cox seemed broadly interested in understanding the platform's role in public life, Rosen had a narrower focus on target setting and problem solving. He was driven, smart, and, in the minds of many, an odd fit for the job.

Rosen had joined Facebook when it acquired his Tel Aviv–based company, Onavo, for $100 million in 2013. Onavo encrypted smart-phone users' activity for privacy purposes—and in the process collected extensive information about their locations, what apps they used, and how long they used them for. Given Zuckerberg's fear of failing to spot upstart competitors before they became entrenched, this was vital business intelligence.

Within months of the purchase, Onavo data detected the explosive growth of messaging service WhatsApp. Though insignificant in the United States, the app was installed on 99 percent of Android phones in Spain and was surging in the developing world. Facebook pounced and bought the service for $19 billion, a shocking price at the time that was later hailed as a bargain.

It was also Onavo data that had detected that Snapchat's "Stories" feature posed an urgent threat. Once Facebook cloned the ephemeral posting format on Instagram in the form of Stories, Onavo data showed the threat had passed, as the *Wall Street Journal* reported in 2017. Rosen's reputation at Facebook was cemented.

If accounts of Onavo's utility took years to become public, there was good reason for it: Facebook didn't want anyone to realize what it was doing. Beyond the ethical questions around a privacy app that harvested data from its users, Onavo's snooping habits were at best a stretch for Google's and Apple's terms of service. Both companies' app stores eventually took down its apps, and the U.S. Federal Trade Commission cited Onavo in its complaint alleging anti-competitive business practices.

That lingering aura of skulduggery made it a little odd when Rosen was picked as the head of Facebook's newly created Integrity division in late 2017. (The company noted that, before his work at Onavo, he'd done security work at Oracle.) Rosen was known for being efficient and had a stellar reputation with the company's leadership, but he reported to Naomi Gleit and the Growth team, which, as Andrew Bosworth had noted in his soon-to-be-infamous "Ugly Truth" memo, was renowned for its ethical flexibility. An unlikely home, in other words, for a group for whom ethics and integrity were supposed to be paramount.

The potential for conflict was clear. Some of the ranking work that McNally's team had already done suggested that increased product usage might be making integrity problems worse. Some integrity work might cut into growth down the line.

Once, when Rosen was presented with data showing how rapidly Russians had managed to establish large-scale accounts on the platform, his response wasn't to question whether there was a flaw in the platform's design. "They're just phenomenal growth marketers," he remarked to a colleague, with a hint of respect. (The company said that such a comment from Rosen would have been sarcasm.) The problem, to Rosen, wasn't that an entity had built up a huge following with vitriolic, spammy content. That was how the platform was supposed to work. The problem was that the entity was Russian.

Rosen thrived on this sort of categorical distinction. Rather than starting with the type of qualitative questions that executives like Gomez-Uribe and McNally had embraced—such as, what were the signatures of trash publishers? How were hyperactive users affecting the platform?—Rosen started from a different place: Was something against Facebook's existing rules *and* quantifiable? If the answer to both was yes, then Rosen, a driven executive, would oversee a relentless campaign to build classifiers capable of detecting the objectionable thing and then grinding it into the dirt. If a problem didn't fit that mold, however, Rosen would tell you to come back when it did.

This classifier- and metrics-driven approach had the advantage of appealing to Facebook's engineering-minded leadership, which liked the idea of building its way out of moderation problems. A disadvantage was that, for all the progress that machine learning had made, Facebook's automated content enforcement systems weren't remotely up to the standards that effective moderation required.

"We do not and possibly never will have a model that captures even a majority of integrity harms, particularly in sensitive areas," one engineer would write, noting that the company's classifiers could identify only 2 percent of prohibited hate speech with enough precision to remove it.

Inaction on the overwhelming majority of content violations was unfortunate, Rosen said, but not a reason to change course. Facebook's bar for removing content was akin to the standard of guilt beyond a reasonable doubt applied in criminal cases. Even limiting a post's distribution should require a preponderance of evidence. The combination of inaccurate systems and a high burden of proof would inherently mean that Facebook generally didn't enforce its own rules against hate, Rosen acknowledged, but that was by design.

"Mark personally values free expression first and foremost and would say this is a feature, not a bug," he wrote.

Publicly, the company declared that it had zero tolerance for hate speech. In practice, however, the company's failure to meaningfully combat it was viewed as unfortunate—but highly tolerable.

If there was a bright side for integrity staffers under Rosen, it was that the Growth team to which their new boss reported unquestionably had leadership's attention. Gleit's orders to steer the company through the aftermath of Cambridge Analytica and Russian interference guaranteed the work would not lack resources. In May 2018, the company had pledged to hire 3,000 new employees and contractors to work on moderation and safety issues. By October, it raised that pledged number to 10,000.

Having failed to get two additional engineers assigned to election issues a year before, Chakrabarti now told the Civic team that senior leadership had said they had a blank check. Some of that money went into building classifiers, since the systems required massive investment to create and thousands of hours of labeling to train. Civic poured $100 million alone into an automated system that could discern whether a post was about politics, government, or societal issues.

Nowhere did that largesse show up so much as in the ranks of Civic's research team. In most Facebook divisions, the ratio of engineers to researchers was generally around twenty to one. At Civic, the ratio was far lower, with anthropologists, sociologists, and political scientists all pursuing various lines of inquiry into the social network's machinations. What made a civic group on the platform successful? they asked. Which communities had the loudest voices? How did misinformation travel through particular demographics? Did user efforts to rebut high-profile falsehoods work?

At home in the United States, these questions had implications for the tenor of public discourse and politics. Abroad, they involved questions of life and death.

Myanmar, ruled by a military junta that exercised near-complete control until 2011, was the sort of place where Facebook was rapidly filling in for the civil society that the government had never allowed to develop. The app offered telecommunications services, real-time news, and opportunities for activism to a society unaccustomed to them.

In 2012, ethnic violence between the country's dominant Buddhist majority and its Rohingya Muslim minority left around two hundred people dead and prompted tens of thousands of people to flee their homes. To many, the dangers posed by Facebook in the situation seemed obvious, including to Aela Callan, a journalist and documentary filmmaker who brought them to the attention of Elliot Schrage in Facebook's Public Policy division in 2013. All the like-minded Myanmar Cassandras received a polite audience in Menlo Park, and little more. Their argument that Myanmar was a tinder-box was validated in 2014, when a hardline Buddhist monk posted a false claim on Facebook that a Rohingya man had raped a Buddhist woman, a provocation that produced clashes, killing two people. But with the exception of Bejar's Compassion Research team and Cox—who was personally interested in Myanmar, privately funding independent news media there as a philanthropic endeavor—nobody at Facebook paid a great deal of attention.

Later accounts of the ignored warnings led many of the company's critics to attribute Facebook's inaction to pure callousness, though interviews with those involved in the cleanup suggest that the root problem was incomprehension. Human rights advocates were telling Facebook not just that its platform would be used to kill people but that it already had. At a time when the company assumed that users would suss out and shut down misinformation without help, however, the information proved difficult to absorb. The version of Facebook that the company's upper ranks knew—a patchwork of their friends, coworkers, family, and interests—couldn't possibly be used as a tool of genocide.

Facebook eventually hired its first Burmese-language content reviewer to cover whatever issues arose in the country of more than 50 million in 2015, and released a packet of flower-themed, peace-promoting digital stickers for Burmese users to slap on hateful posts. (The company would later note that the stickers had emerged from discussions with nonprofits and were "widely celebrated by civil society groups at the time.") At the same time, it cut deals with telecommunications providers to provide Burmese users with Facebook access free of charge.

The first wave of ethnic cleansing began later that same year, with leaders of the country's military announcing on Facebook that they would be "solving the problem" of the country's Muslim minority. A second wave of violence followed and, in the end, 25,000 people were killed by the military and Buddhist vigilante groups, 700,000 were forced to flee their homes, and thousands more were raped and injured. The UN branded the violence a genocide.

Facebook still wasn't responding. On its own authority, Gomez-Uribe's News Feed Integrity team began collecting examples of the platform giving massive distribution to statements inciting violence. Even without Burmese-language skills, it wasn't difficult. The torrent of anti-Rohingya hate and falsehoods from the Burmese military, government shills, and firebrand monks was not just overwhelming but overwhelmingly successful.

This was exploratory work, not on the Integrity Ranking team's half-year roadmap. When Gomez-Uribe, along with McNally and others, pushed to reassign staff to better grasp the scope of Facebook's problem in Myanmar, they were shot down.

"We were told no," Gomez-Uribe recalled. "It was clear that leadership didn't want to understand it more deeply."

That changed, as it so often did, when Facebook's role in the problem became public. A couple of weeks after the worst violence broke out, an international human rights organization condemned Facebook for inaction. Within seventy-two hours, Gomez-Uribe's team was urgently asked to figure out what was going on.

When it was all over, Facebook's negligence was clear. A UN report declared that "the response of Facebook has been slow and ineffective," and an external human rights consultant that Facebook hired eventually concluded that the platform "has become a means for those seeking to spread hate and cause harm."

In a series of apologies, the company acknowledged that it had been asleep at the wheel and pledged to hire more staffers capable of speaking Burmese. Left unsaid was why the company screwed up. The truth was that it had no idea what was happening on its platform in most countries.

Facebook had not invested in people who could speak the world's various languages, and that was compounded by the company's system of classifiers. For some problems, like nudity or gore, Facebook could use the same classifier worldwide—dick pics and decapitations are, alas, universal. But for socially critical problems like hate speech and incitement, the algorithms didn't translate. For every language and dialect that Facebook wanted to succeed at, it would need to build separate algorithms.

With more than 7,000 recognized languages worldwide, the math was not on Facebook's side. Even if the company wrote off anything spoken by fewer than 10 million people, that left more than one hundred languages on the table. Regardless of Facebook's resources, building classifiers for that many languages—let alone good classifiers—was out of the question. Each new language required deep cultural context, machine learning, and regular updating. When Facebook's Safety team was approached by human rights activists warning of coming disaster in Myanmar, it wasn't that they didn't believe them. It was that they were being warned of impending disaster in dozens of countries the company didn't know much about. They couldn't tell the real ones from the false alarms.

McNally, who arrived at the company after the first wave of ethnic cleansing had ended and just before the second one commenced, described it as a failure of understanding. "Facebook did in Myanmar the same thing it does in every other country—whatever the model thinks will be liked and shared gets fed back to the users," McNally said. "What was different is the people pushing misinformation were the establishment. If the majority decides to persecute the minority on a platform based on popularity, the majority wins."

The job of heading off further violence was split between McNally's News Feed Integrity team and the Civic team, which assigned a dedicated staff to the job of cleaning up the platform in Myanmar and, ideally, averting the next such incident elsewhere. Chakrabarti told his team that his instructions from leadership boiled down to "Do what you need to do."

The system that Civic inherited was embarrassing. The classifiers required to predict what content supported political violence were rough or nonexistent, and a technical weakness in Facebook's Burmese user-reporting system meant that reviewers could see posts flagged for hate speech only after a twenty-four-hour delay. When the content did get reviewed, it was often by people who didn't understand it. "Doing Google translate on Burmese is illegible," remarked one person involved.

Civic got to work. They began to systematically suppress posts from users who had repeatedly violated Facebook's community standards, and built a system that prevented questionable users from appearing in friend recommendations. They lowered the level of confidence that Facebook's classifiers needed to achieve before removing content, and started building "Green Lantern," a self-training algorithm that identified variations on slurs and words that appeared to be used as synonyms for prohibited language.

Facebook took one final action. It cut off the telecom subsidies that made Facebook use in Myanmar free. It was as close as the company had ever gotten to acknowledging that its product might not always be worth the societal cost.

Facebook had the sense to know that Myanmar was unlikely to be a one-off and that it would need these sorts of tools again. At the urging of Chakrabarti and McNally, the company set up a new team to handle "At Risk Countries," or ARC. Nobody ever specified what they were at risk of, but the criteria made it clear. To get "At Risk" status, a country had to have a history of violence, a potential trigger for future conflict such as an upcoming election, and a high Facebook market penetration. In other words, the status was reserved for places where Facebook's products could plausibly cause or exacerbate a genocide or civil war.

Designating a country as "At Risk" gave it a shot at receiving monitoring, safety protections, and engineering work beyond what its strategic importance to Facebook could justify. From a business perspective, Facebook had no reason to pay attention to places like Myanmar or Sri Lanka, but they loomed larger from the viewpoint of both basic humanity and potential reputational damage.

The team within Civic that was assigned to ARC was split between those who sought to study and predict the circumstances in which Facebook would be misused and those whose job it was to design tools to prevent that from happening. The team was never huge—a few dozen staffers at its peak—but it would end up playing an outsized role in building the company's broader defenses against misinformation and hate speech. Called in to address crisis after crisis, the team was able to experiment with interventions outside Facebook's normal playbook.

Over time, they built up a library of dozens of prepackaged changes that would come to be known as "Break the Glass" measures, with technical names like "PL7," "PE2," and "P50 Sigmoid Demotion." Some were simple tweaks that imposed stricter rules than those that normally governed the platform. Others were kill switches for specific product features that were known to have chronic integrity problems, such as group recommendations. The most aggressive of the interventions fundamentally altered Facebook's mechanics in ways that the company had never deployed in even crisis situations. (A non-hypothetical example: should you ever log onto the platform to find that Facebook has intentionally disabled News Feed content except for posts from your immediate friends, it's a decent bet that a lot of people are going to die.)

The ARC team's greatest strength stemmed from the fact that nobody really cared what it did. The countries it focused on were seen as all but irrelevant politically and commercially, and the Public Policy team's concern with them was generally limited to the wish that they not cause problems. "Joel cared about four places: the U.S., Europe, India, and Israel," said a former member of his team. (A company spokesman denied this characterization, stating that Kaplan traveled broadly overseas.)

Consequently, issues that Facebook approached as a matter

of absolute principle in developed markets became questions of pragmatism elsewhere. Automatically closing hate-filled comment threads? Not a chance in the U.S. But, in a pinch, it could be temporarily rolled out ahead of Kenya's elections. In Myanmar, the company eventually established a fact-checking program, but it also gave a senior Public Policy executive in Singapore the authority to unilaterally strike down content that seemed dubious and inflammatory—an approach that was far more effective than waiting for outsourced partners to review a questionable post, explain why it was false, and submit it through Facebook's systems. It was also an obvious violation of Zuckerberg's longstanding demand that employees not determine the truth of content on Facebook.

The extra latitude granted to the ARC team had its limits. Zuckerberg and the News Feed Ranking team were adamant that "At Risk" status be temporary. The company wasn't willing to bifurcate its systems on an ongoing basis. And even with "At Risk Countries," the most effective interventions were held in reserve for periods of unusually high risk. They were small markets, but they were still, in the end, markets.

With the ARC team's eyes on the rest of the world, James Barnes was looking closer to home. After liaising with the Trump campaign, he had worked for six months on Facebook's Advocacy team, created at the beginning of 2017, to "use Facebook to tell the Facebook story." With questions about Facebook's role in the 2016 election dragging on its reputation, the company was going in for some soft lobbying.

Barnes was put in charge of "meme busting"—that is, combating the spread of viral hoaxes about Facebook, on Facebook. No, the company was not going to claim permanent rights to all your photos unless you reshared a post warning of the threat. And no, Zuckerberg was not giving away money to the people who reshared a post saying so. Suppressing these digital chain letters had an obvious payoff; they tarred Facebook's reputation and served no purpose.

Unfortunately, restricting the distribution of this junk via News Feed wasn't enough to sink it. The posts also spread via Messenger,

in large part because the messaging platform was prodding recipi-
ents of the messages to forward them on to a list of their friends.

The Advocacy team that Barnes had worked on sat within Facebook's Growth division, and Barnes knew the guy who oversaw
Messenger forwarding. Armed with data showing that the current
forwarding feature was flooding the platform with anti-Facebook
crap, he arranged a meeting.

Barnes's colleague heard him out, then raised an objection.

"It's really helping us with our goals," the man said of the forwarding feature, which allowed users to reshare a message to a list
of their friends with just a single tap. Messenger's Growth staff had
been tasked with boosting the number of "sends" that occurred each
day. They had designed the forwarding feature to encourage precisely the impulsive sharing that Barnes's team was trying to stop.

Barnes hadn't so much lost a fight over Messenger forwarding
as failed to even start one. At a time when the company was trying to control damage to its reputation, it was also being intentionally agnostic about whether its own users were slandering it. What
was important was that they shared their slander via a Facebook
product.

"The goal was in itself a sacred thing that couldn't be questioned,"
Barnes said. "They'd specifically created this flow to maximize the
number of times that people would send messages. It was a Ferrari,
a machine designed for one thing: infinite scroll."

Barnes left that team after a few months, moving to a quiet section
of the company's ad business while the FBI and other various government entities continued to ask him questions about his work on
the Trump campaign.

Following the tumult of the past couple of years, Barnes had
given himself a private ultimatum: he needed to work on efforts to
improve Facebook ahead of the 2018 midterm elections—or quit.
Already friendly with Chakrabarti, Barnes asked if he could join the
Civic team as its first dedicated product manager. Chakrabarti took
him on board.

One of Barnes's early projects was studying how to prevent civic hoaxes from going viral. Suppressing links to known, fact-checked falsehoods was straightforward, but somehow the countermeasures weren't working. An analyst on Barnes's team figured out why: they were thriving via Messenger forwards, just like the Facebook hoaxes. Viral misinformation didn't require the help of Facebook's recommendations to blow up. All it took was for Facebook to lower the bar for sharing it enough that unsavvy users would spam their friends.

Barnes undertook a second effort to kill message forwarding, this time in the name of a well-informed public. He and Chakrabarti got in touch with Site Integrity, the team of engineers that fought porn accounts and fake Ray-Ban ads, and argued that Messenger forwarding was contributing to spam, a problem Site Integrity did not take lightly (a Workplace group devoted to spam fighting was titled "Kill It With Fire").

Although Site Integrity was in no position to kill off a Messenger feature, they did have the authority to cap how heavily it was used, limiting how much harm prolific fake-news sharers could do. The two teams' engineers jointly wrote the code, tested it to prove it cut back on spam, and then shipped it.

The Messenger Growth crew, very intentionally, was not consulted. "We changed their product without telling them," Barnes recalled, smirking.

A couple of days later, Messenger Growth staffers called Barnes and Kaushik Iyer, the head of Civic Engineering and Samidh Chakrabarti's counterpart, into a conference room. They had figured out what Civic had done and it was damaging their usage metrics. They demanded that Civic roll back the message send limits.

Iyer, a soft-spoken presence, said he understood the Messenger team's concerns. But Civic had concluded that unfettered message forwarding created both a spam problem and a threat to the democratic process, so regrettably the rate limit would stay in place.

"If you want to have a Zuck review, I'll arrange time on his calendar," said Iyer, who after eight years on Facebook's Engineering team could credibly claim to set up a brief meeting for the CEO to adjudicate the dispute.

It was less an offer than a challenge. If the Messenger staffers forced the meeting, they would have to argue in favor of allowing messages that the Site Integrity team had already determined were spam, an admission that the team had encouraged trash messages to game their usage targets.

The Messenger Growth team folded. They would have to find another way to boost "sends."

The gutsy maneuver made for the type of story that Facebook engineers across the company might chortle about over a beer. Cramming the work of an ever-growing number of Product and Engineering teams into a single app guaranteed plenty of zero-sum squabbling, and Iyer had just outflanked his opponents. What set this version of the story apart was that Civic wasn't usually the winning team in conflicts against Growth. The victory owed a great deal to the support of Site Integrity, but it was still a victory.

8

Katie Harbath, over on the Politics and Government team, was prepping for the 2018 midterms, and she had plenty on her plate. Her Elections team had narrowly avoided being disbanded in the wake of 2016, and it was no longer offering the same support to campaigns that had earned the credit and the blame for Trump's victory. With a staff of sixty, it now was responsible for making, explaining, and troubleshooting the platform's rules for campaigns.

Harbath was spending ever more time liaising with Civic. She took a particular interest in CORGI, a system the team and Core Data Science had been building since the second half of 2017 to identify accounts that were either working in close coordination or being managed under central control.

CORGI (the "C" stands for "coordinated" and the other letters for things that even some of its most devoted users can't recall) was premised on the idea that, no matter how sophisticated, attempts to manipulate the platform invariably showed their hand. Sometimes the tells were obvious, such as when hundreds of like-minded accounts shared the same IP address and had been created in the span of thirty seconds. Other times, the slipups were subtle—the account interactions a little too reliable, the wording of their posts a little too similar. CORGI was the cute name for an automated way to detect those patterns.

The tool was sufficiently novel and valuable that Facebook's counterespionage teams used it before building their own methods for detecting foreign inauthentic behavior. But Civic was intent on a much broader use. They wanted to ferret out large-scale domestic coordination attempts, too.

In principle, this was not controversial. Facebook's rules prohibited publishers from boosting their content through manipulative tactics, including spamming, fake accounts, and coordinated posting. The definition of "inauthentic behavior" was broad, encompassing anything that appeared geared toward gaining preferential treatment. If the company detected sustained efforts to manipulate the platform or game user attention, the entity behind it was eligible for a permanent ban.

In practice, however, getting a political account network taken down for anything other than efforts at foreign manipulation was a great deal more fraught. Part of the problem was that, by the time Facebook started paying attention to the inauthentic behavior, it was already endemic.

Entities like Liftable Media, a digital media company run by longtime Republican operative Floyd Brown, had built an empire on pages that began by spewing upbeat clickbait, then pivoted to supporting Trump ahead of the 2016 election. To compound its growth, Liftable began buying up other spammy political Facebook pages with names like "Trump Truck," "Patriot Update," and "Conservative Byte," running its content through them.

In the old world of media, the strategy of managing loads of interchangeable websites and Facebook pages wouldn't make sense. For both economies of scale and to build a brand, print and video publishers targeted each audience through a single channel. (The publisher of *Cat Fancy* might expand into *Bird Fancy*, but was unlikely to cannibalize its audience by creating a near-duplicate magazine called *Cat Enthusiast*.)

That was old media, though. On Facebook, flooding the zone with competing pages made sense because of some algorithmic quirks. First, the algorithm favored variety. To prevent a single popular and prolific content producer from dominating users' feeds, Facebook blocked any publisher from appearing too frequently. Running dozens of near-duplicate pages sidestepped that, giving the same content more bites at the apple.

Coordinating a network of pages provided a second, greater benefit. It fooled a News Feed feature that promoted virality. News Feed had been designed to favor content that appeared to be emerging organically in many places. If multiple entities you followed were all talking about something, the odds were that you would be interested, too, so Facebook would give that content a big boost.

The feature played right into the hands of motivated publishers. By recommending that users who followed one page like its near doppelgängers, a publisher could create overlapping audiences, using a dozen or more pages to synthetically mimic a hot story popping up everywhere at once.

Plenty of people at Facebook had understood that digital publishers were playing tricks, but nobody understood how devastatingly effective the networks of cutout pages were. That wouldn't become clear until a junior data scientist named Sophie Zhang later found that the tactic was rampant and appeared to trick News Feed into increasing the distribution of content exponentially.

Zhang, working on the issue in 2020, found that the tactic was being used to benefit publishers (*Business Insider, Daily Wire,* a site named iHeartDogs), as well as political figures and just about anyone interested in gaming Facebook content distribution (Dairy Queen franchises in Thailand). Outsmarting Facebook didn't require subterfuge. You could win a boost for your content by running it on ten different pages that were all administered by the same account.

It would be difficult to overstate the size of the blind spot that Zhang exposed when she found it. "Should we not be looking at whether we can adjust?" David Agranovich, the company's director of Global Threat Disruption and a former National Security Council director for intelligence, wrote in response to Zhang's findings. Zhang, who was largely moonlighting on Civic matters, would later be fired for paying inadequate attention to her day job while she focused on the platform's vulnerabilities to political manipulation.

Even if Facebook was still years away from understanding the force multiplier that networked accounts offered, by the time that the U.S. midterms approached in 2018, the company did, at least, understand they were trouble. Whether due to principle, pride, or ineptitude, conventionally respectable news outlets generally hadn't played these distribution games, or at least not well. That had left the field open for a new breed of Facebook-optimized publishers: ones willing to do anything the platform rewarded for views.

Liftable was an archetype of that malleability. The company had begun as a vaguely Christian publisher of the low-calorie inspirational content that once thrived on Facebook. But News Feed was a fickle master, and by 2015 Facebook had changed its recommendations in ways that stopped rewarding things like "You Won't Believe Your Eyes When You See This Phenomenally Festive Christmas Light Show."

The algorithm changes sent an entire class of rival publishers like Upworthy and ViralNova into a terminal tailspin, but Liftable was a survivor. In addition to shifting toward stories with headlines like "Parents Furious: WATCH What Teacher Did to Autistic Son on Stage in Front of EVERYONE," Liftable acquired WesternJournal .com and every large political Facebook page it could get its hands on.

This approach was hardly a secret. Despite Facebook rules prohibiting the sale of pages, Liftable issued press releases about its acquisition of "new assets"—Facebook pages with millions of followers. Once brought into the fold, the network of pages would blast out the same content.

Nobody inside or outside Facebook paid much attention to the craven amplification tactics and dubious content that publishers such as Liftable were adopting. Headlines like "The Sodomites Are Aiming for Your Kids" seemed more ridiculous than problematic. But Floyd and the publishers of such content knew what they were doing, and they capitalized on Facebook's inattention and indifference.

"The smart ones built pages early on, in the 2013 timeframe, then they started merging and monetizing them," Harbath said.

By 2017, when Facebook started worrying about problems like misinformation, engagement bait, and coordinated posting, partisan publishers—mostly on the right, but some on the left—were well established. Liftable alone was a notable player in the conservative news business, bringing in a billion page views a year and eight-digit ad revenues. It was a little late to shut the barn door.

The problem was like many at Facebook, in that it stemmed from inaction on what had been only a minor concern a few years earlier. The company's failure to deal with problems hadn't just created bad precedents—it had created bad ecosystems.

The early work trying to figure out how to police publishers' tactics had come from staffers attached to News Feed, but that team was broken up during the consolidation of integrity work under Guy Rosen, who was focused on battling harmful content rather than the abuses that helped it spread.

"The News Feed integrity staffers were told not to work on this, that it wasn't worth their time," recalled product manager Elise Liu.

In contrast, Liu and Sagnik Ghosh, her engineering partner at Civic, viewed it as a battle that the platform couldn't afford to lose. Facebook promoted content depending on whether networks of users shared it; allowing small groups of people controlling large groups of dummy accounts to fake that buzz would artificially boost their content above legitimately shared posts. So Civic picked up where the News Feed team had left off.

Facebook's policies certainly made it seem like removing networks of fake accounts shouldn't have been a big deal: the platform required users to go by their real names in the interests of accountability and safety. In practice, however, the rule that users were allowed a single account bearing their legal name generally went unenforced.

There were valid reasons why a user might not abide by it—maybe they'd lost access to their profile when they broke their phone, or maybe they lived in a place where supporting democracy leads to a late-night knock on the door. But Facebook didn't require what it called SUMAs—"same user, multiple accounts"—to justify themselves. The company wasn't in the business of keeping new users off its platform, even when they weren't strictly new.

Civic's plan to track and target networks of fake accounts was therefore getting into novel territory. Russians were forbidden from pretending to be Americans in the interest of manipulating American politics; what Facebook's leadership thought of Americans using fake accounts for an identical purpose wasn't certain. Still, clear rules banning fake accounts were on the books, and nobody said they couldn't be applied to trash partisan publishers.

"Facebook can be a very bottom-up company," said Ghosh. "If Mark had his eye on this stuff, it's not clear this would have happened."

The job of explaining the new ground rules to entities like Floyd Brown's fell to Harbath, who, as a prominent Republican for Facebook in Washington, DC, sometimes drew the role of walking partisan publishers through the platform's new rules of the road, a process that was often acrimonious. Publishers didn't like being told that they would need to stick closer to the facts and stop promoting their content with throwaway pages.

In March 2018, Brown went on Tucker Carlson's Fox News show to accuse Facebook of being the "thought police" as it began to crack down on these sorts of schemes, painting it as an anti-conservative attempt to control the discourse ahead of the midterms.

For all the tirades, Liftable cleaned up its act, deleting false news stories and rebranding or merging its scores of pages under the "Western Journal" mantle, for a fresh start. "Western Journal" was still amplifying its content in ways that were officially frowned upon, but Facebook considered the company to have come close enough and let it be.

Not every publisher was willing to play ball. In the spring of 2018, the Civic team began agitating to address dozens of other networks of recalcitrant pages, including one tied to a site called "Right Wing News." The network was run by Brian Kolfage, a U.S. veteran who had lost both legs and a hand to a missile in Iraq.

Harbath's first reaction to Civic's efforts to take down a prominent disabled veteran's political media business was a flat no. She couldn't dispute the details of his misbehavior—Kolfage was using fake or borrowed accounts to spam Facebook with links to vitriolic, sometimes false content. But she also wasn't ready to shut him down for doing things that the platform had tacitly allowed.

"Facebook had let this guy build up a business using shady-ass tactics and scammy behavior, so there was some reluctance to basically say, like, 'Sorry, the things that you've done every day for the last several years are no longer acceptable,'" she said.

As she had with Floyd Brown, Harbath began speaking directly with Kolfage about how to clean up his operation. "Early on, there was the plausible scenario that these guys didn't know the rules," Harbath said. "I had a big concern of not having things super tight before taking down the pages of a triple amputee Iraq war vet."

The conversations were difficult.

"He's bombastic, but so are a lot of people in my inbox," Harbath said.

She thought she was making progress, until the Civic team told her that Kolfage was both failing to reform his tactics (the administrators of a single group tied to Kolfage were running 189 pages, 70 of which existed primarily to post duplicate links to one of his news sites) and even mocking Harbath's reform efforts via direct messages with his accomplices.

She tried again, getting on a conference call with a colleague in Content Policy and a lawyer for Kolfage to stress that the company was serious. That didn't work either. Rather than immediately take Kolfage's network down, Facebook first imposed some modest penalties as a warning shot.

All that accomplished was pissing Kolfage off. His emails to Harbath, which she shared with Civic staffers at the time, weren't so much rude as menacing. "We were sending analyses to Katie, she was sending them to Kolfage, and he was responding with threats," Liu recalls.

Other than simply giving up on enforcing Facebook's rules, there wasn't much left to try. Facebook's Public Policy team remained uncomfortable with taking down a major domestic publisher for inauthentic amplification, and it made the Civic team prove that Kolfage's content, in addition to his tactics, was objectionable. This hurdle became a permanent but undisclosed change in policy: cheating to manipulate Facebook's algorithm wasn't enough to get you kicked off the platform—you had to be promoting something bad, too.

Facebook decided to let Civic go ahead with its takedown of pages such as Kolfage's. But the midterms were just a few weeks away, and they expected blowback. So the company's public relations teams asked that Civic go through with any takedowns it was planning to do before the election *in one fell swoop.* This would hit both left- and right-leaning publishers, giving the company its much-desired veneer of "balance." If Facebook was going to tangle with political publishers, it might as well get it over with.

On the morning of October 11, 2018, the team behind the take-downs flipped the switch and nuked more than eight hundred accounts and pages catering to liberals and conservatives, taking a photo to commemorate the occasion. Nathaniel Gleicher, head of Cybersecurity Policy, announced the removals, calling them a nec-essary response to publisher deception.

True to form, Kolfage didn't go quietly. "I've seen my legs blown to pieces and lost friends," Kolfage wrote on "Right Wing News," elic-iting readers' sympathy. He accused Harbath by name of destroying a wounded veteran's livelihood, falsely claiming that she had "never once" raised concerns about his pages. Liberal publishers who had their pages removed also declared themselves baffled.

Fighting with established Facebook publishers didn't win the company any friends. Conservative outlets including Fox News broadly accepted Kolfage's accusations of anti-conservative bias, while outlets such as the *Guardian* and the *Washington Post* wrote about bipartisan "accusations of censorship."

Kaplan, who had given Harbath the go-ahead to proceed with the takedown, asked her whether the removals had truly needed to happen.

"Yes, Joel, I do think it was the right call," Harbath responded. "We gave them every opportunity."

Kaplan looked skeptical, but he didn't push the matter. (The deci-sion to remove Kolfage in particular would look even better when he pled guilty to fraud charges stemming from his collection of $25 million, much of it on Facebook, to privately fund construction of Trump's border wall. He was eventually sentenced to more than four years in prison.)

Whatever mixed feelings there might have been about the effort elsewhere at Facebook, the Civic team was jubilant. Tests showed that the takedowns cut the amount of American political spam content by 20 percent overnight. Chakrabarti later admitted to his subordinates that he had been surprised that they had succeeded in taking a major action on domestic attempts to manipulate the platform. He had privately been expecting Facebook's leadership to shut the effort down.

"That massive takedown demonstrated that you could meaningfully alter the flow of content on the platform in ways that were healthy and get approval," said Liu, the Civic product manager. "And yes, there were a few bruises, but everyone was okay at the end."

Though Facebook had made progress, both Civic and senior executives remained jittery as the midterms approached. Facebook couldn't afford to screw up its handling of another American election, even as it was still mired in controversies from the last one.

A month and a half before Election Day, Civic launched a "War Room" to carry it through. Whether the initiative was envisioned as a way to marshal resources in a central location or a public relations gimmick is a matter of dispute, but the final result combined both. Reporters from the *New York Times, The Verge,* and *TechCrunch* took tours and wrote about the twenty-odd staffers who had done up a conference room at Menlo Park with an American flag and oversized monitors displaying dashboards. Chakrabarti talked up the company's improved enforcement tools, its deep bench of security experts, and its efforts to game out threats.

Whatever public reassurance the War Room was meant to provide, it wasn't just a Potemkin village. The real-time monitoring systems were genuine, and so was the operation's capacity to escalate problems to the company's leadership. Every night, a member of the team sent a summary of the day's events to Zuckerberg, Sandberg, and other executives detailing what they had accomplished, what they needed help with, and what they were worried about.

In a matter of a few months, Civic had scoured the company's various Integrity teams for new "Break the Glass" tools beyond what the At Risk Countries team had put together. Civic found that a team building hate speech classifiers had been working on a plan to freeze comments on "hate bait"—that is, racially or religiously charged posts that thrived on the engagement of users who filled their comments sections with slurs. But the Public Policy team had blocked the proposal on the grounds that it wasn't fair to penalize a post because people left bigoted comments.

Civic resurrected the proposal and submitted it to Sandberg, who approved it within a day. The team then drew up plans to stop recommending groups during volatile times and prepared a true circuit breaker. If conditions got hairy enough, the company could turn off posts from publishers, pages, and groups—everything but content from a user's friends. The company tested the maneuver, not to find out if it would tank the company's usage metrics—of course it did—but to make sure that disabling so much of the platform's mechanics at once wouldn't crash it.

Trying to prepare for every conceivable disaster was a white-knuckle effort, one that produced a spreadsheet of emergency levers to tame the platform alongside the contacts of engineers across the company that could pull them at a moment's notice. The around-the-clock activity also produced a sense of camaraderie. Some of the younger members of the team who didn't have families decided that, after spending seven days a week in the office for the campaign's final stretch, going home was too much work. Rather than book-ing hotel rooms near the office, Liu put $20,000 on a credit card for three weeks' rent on a mansion in nearby Atherton. Barnes was among the staffers who took up residence there. The work felt good to him, like a chance to help fix an error.

—

Civic had one last job to take care of before they could turn all their attention to the U.S. midterms. Brazil, the world's second-largest democracy, was due to hold a presidential election in October. It was a brawl between an inflammatory conservative populist and an established politician, just like in the U.S. two years prior.

A few weeks before Brazilians went to the polls, Chakrabarti told Cox and Zuckerberg to stop by the War Room to see what the team was up to. Zuckerberg poked his head in for a few minutes one day and then ducked back out without saying much. Cox, however, dropped by unannounced one afternoon and began asking the staffers about what they were seeing in Brazil and what worried them most.

There was a reason that Cox gave a welcome talk to nearly every batch of incoming employees. More than anyone else in Facebook's upper ranks, the head of Product had a gift when it came to putting people at ease and making them feel comfortable speaking up. After asking questions, Cox would look people in the eye and tell them to write to him personally if there was something they needed. The guy clearly cared—about employees, about Facebook's mission, and about the world in general.

Cox was in fine form the day he stopped by the War Room, spending ninety minutes with the Civic and Integrity staffers on duty. One of Harbath's staffers present alerted her to what Harbath would later call "Cox's Reverse Q&A." Out on the East Coast, Harbath rushed home to watch in horror as Cox coaxed a group of wired and stressed Civic staffers into venting their frustrations—many of which were with Facebook's Public Policy team.

A staffer had shown Cox that a Brazilian legislator who supported the populist Jair Bolsonaro had posted a fabricated video of a voting machine that had supposedly been rigged in favor of his opponent. The doctored footage had already been debunked by fact-checkers, which normally would have provided grounds to bring the distribution of the post to an abrupt halt. But Facebook's Public Policy team had long ago determined, after a healthy amount of discussion regarding the rule's application to President Donald Trump, that government officials' posts were immune from fact-checks. Facebook was therefore allowing false material that undermined Brazilians' trust in democracy to spread unimpeded.

Cox was troubled. The next day he began setting up meetings with Kaplan, Sandberg, and other senior executives. He also asked Chakrabarti to make the case for reconsidering politicians' exemption from fact-checking during Sandberg's weekly Friday meeting with senior executives.

The day of the meeting, Harbath conferenced in from Washington. Sandberg told everyone they had half an hour to sort the question out. Chakrabarti began, arguing that, despite the company's reluctance to police political speech, it had to draw the line at efforts to undermine elections. Such lies were toxic to democracy, *especially* when they came from politicians.

Both Sandberg and Cox were nodding.

"Sheryl was super close to saying, 'We're going to reverse this,'" Harbath recalled.

Then Facebook's Washington staff asked everyone in the room but Cox and Sandberg to clear out for a few minutes so they could make their case. Chakrabarti and others sat outside the glass-walled conference room—a space jokingly named "Only Good News"—while executives discussed the risk that changing policy shortly before two major elections would put Facebook in the spotlight at a time when the company should be keeping its head down. Harbath rang a member of her team in Brazil, who heartily concurred.

After a few minutes, Sandberg called Chakrabarti and the rest of the executives back in. Both she and Cox now firmly agreed that the idea should be tabled, at least until after the Brazil and U.S. elections.

Despite Civic's concerns, voting in Brazil went smoothly. The same couldn't be said for Civic's colleagues over at WhatsApp. In the final days of the Brazilian election, viral misinformation transmitted by unfettered forwarding had blown up. The potential for such abuse was exactly the sort that had motivated Barnes and Iyer to pick a fight with the Messenger team by imposing rate limits on messaging. Civic didn't feel great about that, but at least their platform wasn't the one in trouble this time.

The celebration was short-lived. Sometime early that evening, everyone's phones began to ping—an automated system set up to detect emerging problems had just alerted that hate speech in Brazil was spiking.

Supporters of the victorious Bolsonaro, who shared their candidate's hostility toward homosexuality, were celebrating on Facebook by posting memes of masked men holding guns and bats. The accompanying Portuguese text combined the phrase "We're going hunting" with a gay slur, and some of the posts encouraged users to join WhatsApp groups supposedly for that violent purpose. Engagement was through the roof, prompting Facebook's systems to spread them even further.

While the company's hate classifiers had been good enough to detect the problem, they weren't reliable enough to automatically remove the torrent of hate. Rather than celebrating the race's conclusion, Civic War Room staff put out an after-hours call for help from Portuguese-speaking colleagues. One polymath data scientist, a non-Brazilian who spoke great Portuguese and happened to be gay, answered the call.

For Civic staffers, an incident like this wasn't a good time, but it wasn't extraordinary, either. They had come to accept that unfortunate things like this popped up on the platform sometimes, especially around election time.

It took a glance at the Portuguese-speaking data scientist to remind Barnes how strange it was that viral horrors had become so routine on Facebook. The volunteer was hard at work just like everyone else, but he was quietly sobbing as he worked. "That moment is embedded in my mind," Barnes said. "He's crying, and it's going to take the Operations team ten hours to clear this."

The U.S. midterms had arrived. On election night, Barnes and some colleagues stayed in the War Room monitoring the platform until the early hours before going back to the rental house to act like normal tech employees after a successful product launch by getting really drunk. A video Liu recorded on her phone shows Barnes sitting on a couch playing guitar while colleagues try to improvise lyrics to ballads about Civic Integrity.

Again, nothing had gone disastrously wrong. It was time for a victory lap. The following day, the house's residents boarded a flight to DC for a post-election debrief, arriving to find that the Policy team had ordered boxes of cupcakes iced to spell out "Thank you for making us safer."

The trip had been Chakrabarti's idea. After a tense election season, the goal was for his staff to spend a bit of time with their colleagues in Washington. Facebook's Civic and Public Policy teams were on a long road, and maybe things would run more smoothly if they got to know each other better. Chakrabarti asked if Kaplan could give a talk to the visiting Civic staffers toward the end of the trip, but the timing hadn't worked out. Instead, Kaplan offered Kevin Martin, his deputy and friend.

Before joining Facebook earlier that year, Martin had worked as a lobbyist and then chairman of the Federal Communications Commission during the George W. Bush administration. His performance overseeing the regulator was sufficiently problematic that it kicked off a bipartisan investigation of his tenure, which found that he had sometimes deceived his fellow commissioners, failed to address misspending by his deputies, and retaliated against staffers who didn't produce work supporting his preferred policy decisions. The report found nothing illegal, but it hardly spoke to his strengths as a manager.

The plan was for Martin to close out Civic's trip with a thank-you and a Q&A. As the Civic team gathered, more than fifty people in all, Martin was nowhere to be found. Harbath took the stage and stalled until he showed up, fifteen minutes late. Martin promptly began to denigrate Civic's work, questioning the need for fact-checking, brushing off difficult questions about Facebook's ad policies, and challenging the value of integrity work overall.

Adding to the insult was that Martin didn't appear to understand either what Civic did or how Facebook's systems worked. A senior Public Policy official wasn't expected to have a nuanced understanding of how to build a classifier, but Martin seemed unfamiliar with the basics. He could not accept that Facebook couldn't always explain the reason that a piece of content got taken down and he didn't seem to have a grasp on the elements of News Feed ranking.

A company spokeswoman would later dispute that Martin's talk was poorly received, though his audience remembered things differently. The meeting was motivating all right, but not in the way Chakrabarti had hoped. Retreating to a bar after it was all over, several Civic staffers spoke heatedly about quitting.

So much for those cupcakes.

Those working on integrity, and those who cared about Facebook's role in polarization more broadly, were in for a big change.

Cox had been with the company for thirteen years when he left in March 2019. It was he who had launched Facebook's initial post-2016 ranking work, sponsored the growth of a formidable Integrity team, and encouraged them to look into ways the platform might not be making the world a better place. And now he was gone. There would be no transition period or emeritus status—Cox would henceforth be spending time with his family and pursuing philanthropic endeavors.

Even people who spoke directly with Cox in the days after he announced his departure aren't fully in agreement about why he chose to leave.

One explanation was frustration with the company itself. Cox had brought in people to study issues like polarization and encouraged them to produce visionary solutions, then heralded their work in high-profile forums. At the 2018 Aspen Ideas Festival, he'd declared that the company had an "immense responsibility" to ensure that high-quality content won out on Facebook and that the company was hard at work to "make the conversation more civil." He pledged Facebook would soon require large pages to prove their true ownership and told the audience to expect Facebook to start penalizing comments that other users downvoted as abrasive.

As the company's head of Product, Cox was theoretically the guy responsible for bringing these efforts to fruition—but they'd been watered down or blocked. That had been tough on him personally, and he privately told some people who asked about his departure that he and Zuckerberg didn't see eye to eye on the platform's social responsibilities. On his way out, he wrote a goodbye post on Workplace declaring that "social media's history is not yet written, and its effects are not neutral."

Facebook was also, in some ways, getting boring. Zuckerberg had recently announced a company-wide strategy shift to focus on interoperable messaging, allowing users of the Facebook app to chat directly with someone on WhatsApp or Instagram. Technically the problem was a bear, and nobody had the slightest idea how Facebook and Instagram could be made interoperable with WhatsApp without breaking the latter's encryption. More generally, though, it was completely uninteresting and something that users hadn't asked for. "Do you have a sense of why Mark is doing this?" Cox asked an only moderately senior WhatsApp executive around the end of 2018.

The most compelling explanation that anyone could come up with, inside or outside Facebook, was that scrambling the company's technical infrastructure was intended as a poison pill for any attempted corporate breakup. Regulators couldn't carve up the world's largest social media company if its platforms were all just veneers on top of the same product.

Both of these explanations for Cox's exit suggested a growing distance, if not a rift, between Facebook's CEO and its chief product officer. But there was a third explanation that Cox gave people. He had begun thinking about when the right time would be to step away from a company that had been so central to his life. He couldn't come up with an answer; there wasn't one. So he quit.

Whatever the reason, Cox's absence was deeply felt—especially when Civic ran up against a wall in India.

India was a huge target for Facebook, which had already been locked out of China, despite much effort by Zuckerberg. The CEO had jogged unmasked through Tiananmen Square as a sign that he wasn't bothered by Beijing's notorious air pollution. He had asked President Xi Jinping, unsuccessfully, to choose a Chinese name for his first child. The company had even worked on a secret tool that would have allowed Beijing to directly censor the posts of Chinese users. All of it was to little avail: Facebook wasn't getting into China. By 2019, Zuckerberg had changed his tune, saying that the company didn't *want* to be there—Facebook's commitment to free expression was incompatible with state repression and censorship. Whatever solace Facebook derived from adopting this moral stance, succeeding in India became all the more vital: If Facebook wasn't the dominant platform in either of the world's two most populous countries, how could it be the world's most important social network?

The Civic team felt prepared as it readied itself for India's general election in April. They had a rough playbook now for high-profile contests, following the success of the U.S. midterms, and they were ready to put it into practice.

If India's 2014 election, which ushered in Hindu nationalist Narendra Modi, was anything to go by, the team had its work cut out for it. During that campaign, Modi's party, the Bharatiya Janata Party, or BJP, had used unsavory if not yet explicitly banned tactics on Facebook, setting up so-called IT cells to astroturf the platform and amplify news sites that party allies covertly paid for.

Although the BJP was widely regarded as the most sophisticated at digital campaigning, the Indian National Congress or INC, the main opposition party, had also set up IT cells that basically functioned as domestic troll farms. And Facebook, with 200 million daily users in India for its flagship app and more than 400 million daily WhatsApp users, was the primary forum.

National-level operatives had enough sense to be slightly discreet, but local political bosses seemed to *want* to tell Facebook's employees what they were doing. They were proud of it. On more than one occasion, party officials treated Facebook employees to tours of cramped offices where scores of staffers working in shifts banged out Facebook posts and comments. In the most surreal instance, a local party operative gave a Facebook employee a tour of a warehouse outfitted with thousands of phones on racks. The phones would light up as software fed them campaign messages character by character, and then buzz in unison when the message was sent. The site was a mass WhatsApp texting operation, in contravention of the app's send limits.

With the fervor of an idealist, Chakrabarti told his team that, at minimum, they were to replicate what they had done in the U.S. and, even better, they could use the Indian election to improve their tactics even more. What he didn't tell them was the likely resistance they would face from Harbath's counterpart in Delhi, Ankhi Das, the head of India's Policy team.

Das had a longstanding relationship with Hindu nationalist politicians dating back to her time as a lobbyist for Microsoft, where she worked until joining Facebook in 2011. During the 2014 campaign, she had visited Modi's headquarters frequently, though this raised few eyebrows. Facebook was excited about any candidate who was excited about Facebook, and Modi was positioning himself as a tech-savvy economic reformer, far from his history as the local head of Gujarat province when it was plagued by deadly anti-Muslim riots.

Civic's work got off to an easy start because the misbehavior was obvious. Taking only perfunctory measures to cover their tracks, all major parties were running networks of inauthentic pages, a clear violation of Facebook rules.

The BJP's IT cell seemed the most successful. The bulk of the coordinated posting could be traced to websites and pages created by Silver Touch, the company that had built Modi's reelection campaign app. With cumulative follower accounts in excess of 10 million, the network hit both of Facebook's agreed-upon standards for removal: they were using banned tricks to boost engagement and violating Facebook content policies by running fabricated, inflammatory quotes that allegedly exposed Modi opponents' affection for rapists and that denigrated Muslims.

With documentation of all parties' bad behavior in hand by early spring, the Civic staffers overseeing the project arranged an hour-long meeting in Menlo Park with Das and Harbath to make the case for a mass takedown. Das showed up forty minutes late and pointedly let the team know that, despite the ample cafés, cafeterias, and snack rooms at the office, she had just gone out for coffee. As the

Civic Team's Liu and Ghosh tried to rush through several months of research showing how the major parties were relying on banned tactics, Das listened impassively, then told them she'd have to approve any action they wanted to take.

The team pushed ahead with preparing to remove the offending pages. Mindful as ever of optics, the team was careful to package a large group of abusive pages together, some from the BJP's network and others from the INC's far less successful effort. With the help of Nathaniel Gleicher's security team, a modest collection of Facebook pages traced to the Pakistani military was thrown in for good measure.

Even with the attempt at balance, the effort soon got bogged down. Higher-ups' enthusiasm for the takedowns was so lacking that Chakrabarti and Harbath had to lobby Kaplan directly before they got approval to move forward.

"I think they thought it was going to be simpler," Harbath said of the Civic team's efforts.

Still, Civic kept pushing. On April 1, less than two weeks before voting was set to begin, Facebook announced that it had taken down more than one thousand pages and groups in separate actions against inauthentic behavior. In a statement, the company named the guilty parties: the Pakistani military, the IT cell of the Indian National Congress, and "individuals associated with an Indian IT firm, Silver Touch."

For anyone who knew what was truly going on, the announcement was suspicious. Of the three parties cited, the pro-BJP propaganda network was by far the largest—and yet the party wasn't being called out like the others.

Harbath and another person familiar with the mass takedown insisted this had nothing to do with favoritism. It was, they said, simply a mess. Where the INC had abysmally failed at subterfuge, making the attribution unavoidable under Facebook's rules, the pro-BJP effort had been run through a contractor. That fig leaf gave the party some measure of deniability, even if it might fall short of plausible.

If the announcement's omission of the BJP wasn't a sop to India's ruling party, what Facebook did next certainly seemed to be. Even as it was publicly mocking the INC for getting caught, the BJP was privately demanding that Facebook reinstate the pages the party claimed it had no connection to. Within days of the takedown, Das and Kaplan's team in Washington were lobbying hard to reinstate several BJP-connected entities that Civic had fought so hard to take down. They won, and some of the BJP pages got restored.

With Civic and Public Policy at odds, the whole messy incident got kicked up to Zuckerberg to hash out. Kaplan argued that applying American campaign standards to India and many other international markets was unwarranted. Besides, no matter what Facebook did, the BJP was overwhelmingly favored to return to power when the election ended in May, and Facebook was seriously pissing it off.

Zuckerberg concurred with Kaplan's qualms. The company should absolutely continue to crack down hard on covert foreign efforts to influence politics, he said, but in domestic politics the line between persuasion and manipulation was far less clear. Perhaps Facebook needed to develop new rules—ones with Public Policy's approval.

The result was a near moratorium on attacking domestically organized inauthentic behavior and political spam. Imminent plans to remove illicitly coordinated Indonesian networks of pages, groups, and accounts ahead of upcoming elections were shut down. Civic's wings were getting clipped.

Chakrabarti later told his team that he considered the decision to back off enforcing Facebook's stated rules against political parties to be indefensible. But there was an India-specific inconsistency that had bothered him, too. If Facebook's newfound reluctance to intervene against domestically coordinated inauthentic behavior had justified the restoration of BJP-friendly shill accounts, he had asked Zuckerberg, shouldn't Facebook roll back its takedowns of the Indian National Congress entities, too?

"No," Zuckerberg responded. "I think that would just cause more problems."

In an after-action summary of the company's 2019 Indian election integrity work, Chakrabarti gave the overall effort—by the company, and by him—a scathing review. Not only had Facebook failed to enforce its own rules; it had failed to treat all parties equally.

That Civic's defeat came on the heels of Cox's departure didn't feel like a coincidence to many in and around Civic. For two years, the executive had been the biggest champion of integrity work, providing Chakrabarti with not just resources but advice on how to shepherd his work through Facebook.

"Samidh viewed Cox as a mentor," Harbath said. "He also thought Chris was the one who would protect him the most."

With Cox gone, Chakrabarti had no high-level interlocutor to carry his message. And, with a tendency to be strident, he ended up irritating a lot of people who outranked him, including Guy Rosen. At the start of the year, Rosen had set up regular meetings in which Chakrabarti and Harbath would apprise a slew of executives, including Sandberg and Kaplan, of upcoming international elections work. After Cox left, the meetings started being canceled and were never rescheduled.

The disregard was increasingly mutual. Chakrabarti stopped doing media appearances on behalf of the company, confiding to colleagues that he wasn't willing to parrot talking points that he no longer believed in.

Though integrity enforcement work across the company had slowed down while Civic tried to win back some of the autonomy it had lost during the India debacle, the team spent the summer of 2019 doing what it could.

At Chakrabarti's urging, the team also adopted a new approach to advocating for its work. Instead of making the case for why Facebook needed to combat abuses on the platform because they harmed users or political discourse, Civic staffers began to argue that the company needed to address "headline risk," a euphemism for the public humiliation of appearing inept.

One example of the genre came in August 2019, when a researcher warned that, as in 2016, many of the platform's most successful American political pages and groups still weren't American. Though Michael McNally's staff and the News Feed Ranking team had driven the platform's original hoax news problem into remission, Facebook's reluctance to embrace additional pro-quality measures had left an opening for a new generation of political spam farms to take off. Branded "Low-Quality Civic Exporters," such operations had metastasized, taking root in Latvia, Pakistan, Vietnam, and anywhere else with an underemployed labor force and a reliable internet connection. U.S. users were consuming 75 percent of Macedonian political group content on Facebook; on Instagram, it was the "high-violation countries" such as Bangladesh and Uganda that were heavy hitters.

The foreign networks were spreading divisive content, pushing scam e-commerce links, and fraudulently soliciting donations on behalf of an American veterans charity. The proposed justification for taking action, however, wasn't that foreigners were abusing Facebook users. It was that foreigners were abusing Facebook users *too obviously.*

"These examples are so blatant and unsophisticated that our failure to act undermines the credibility of our integrity efforts," the researcher wrote in a memo, declaring the fact that the abuse could "easily be detected externally" as a reason "why it matters."

A company spokeswoman would later dispute that the tactic was necessary, saying that "product and policy decisions are made on merits" at the company. But former Civic staffers would later say that reframing abuse problems as PR concerns won them a few victories: nobody on Facebook's Communications team relished the idea of explaining why Facebook hadn't noticed something fishy about a massive group named "News of the United State" [*sic*]. But the need for the tactic reflected how weak Civic had become inside Facebook—and how little the team trusted Facebook's leadership to do the right thing.

9

Kang-Xing Jin was an unlikely dissident.

He had been friends with Zuckerberg since their first day of classes at Harvard back in 2002. "I was late so I threw on a T-shirt and didn't realize until afterwards it was inside out and backwards with my tag sticking out the front," Zuckerberg reminisced when he returned to Harvard in 2017 to speak at commencement and collect an honorary degree. "I couldn't figure out why no one would talk to me—except one guy, KX Jin, he just went with it." They would go on to share computer science problem sets and late-night meals at Pinocchio's Pizza, better known to Cambridge denizens as Noch's. It was there, per a Facebook post Zuckerberg wrote years later, that they talked about "how one day someone was going to build a community to connect the whole world."

Jin was there, too, the night Zuckerberg launched Facebook from his dorm room. Out of Facebook's more than 2 billion users, he was the sixteenth.

When Zuckerberg and his roommate Dustin Moskovitz dropped out of Harvard after two years to turn Facebook into a business, Jin stayed behind and went on to graduate with honors. He joined the company in the summer of 2006 as part of the original team that launched News Feed and then rose to vice president of Engineering. For Jin's tenth anniversary at the company, Zuckerberg presented him with a wooden replica of a Noch's takeout pizza box, declaring that "over the years he's worked on almost everything we've built."

Jin stood out for both his pre-college friendship with Zuckerberg and his loyalty. As early Facebookers drifted off with pockets of cash after Facebook's 2012 IPO, he stayed put, becoming one of the company's longest-tenured employees.

By 2019, Jin's standing inside the company was slipping. He had made a conscious decision to stop working so much, offloading parts of his job onto others, something that did not conform to Facebook's culture. More than that, Jin had a habit of framing what the company did in moral terms. Was this good for users? Was Facebook truly making its products better?

Other executives were careful when bringing decisions to Zuckerberg to not frame decisions in terms of right or wrong. Everyone was trying to work collaboratively, to make a better product, and whatever Zuckerberg decided was good. Jin's proposals didn't carry that tone. He was unfailingly respectful, but he was also clear on what he considered the range of acceptable positions. Alex Schultz, the company's chief marketing officer, once remarked to a colleague that the problem with Jin was that he made Zuckerberg feel like shit.

In July 2019, Jin wrote a memo titled "Virality Reduction as an Integrity Strategy" and posted it in a 4,200-person Workplace group for employees working on integrity problems. "There's a growing set of research showing that some viral channels are used for bad more than they are used for good," the memo began. "What should our principles be around how we approach this?" Jin went on to list, with voluminous links to internal research, how Facebook's products routinely garnered higher growth rates at the expense of content quality and user safety. Features that produced marginal usage increases were disproportionately responsible for spam on WhatsApp, the explosive growth of hate groups, and the spread of false news stories via reshares, he wrote.

None of the examples were new. Each of them had been previously cited by Product and Research teams as discrete problems that would require either a design fix or extra enforcement. But Jin was framing them differently. In his telling, they were the inexorable result of Facebook's efforts to speed up and grow the platform.

The response from colleagues was enthusiastic. "Virality is the goal of tenacious bad actors distributing malicious content," wrote one researcher. "Totally on board for this," wrote another, who noted that virality helped inflame anti-Muslim sentiment in Sri Lanka after a terrorist attack. "This is 100% direction to go," Brandon Silverman of CrowdTangle wrote.

After more than fifty overwhelmingly positive comments, Jin ran into an objection from Jon Hegeman, the executive at News Feed who by then had been promoted to head of the team. Yes, Jin was probably right that viral content was disproportionately worse than nonviral content, Hegeman wrote, but that didn't mean that the stuff was bad *on average.*

Jin replied by saying that it didn't matter that, on average, the content was okay. The problem was that Facebook's products were making the content worse. What wasn't objectionable tended to be inane. Surveys, Jin noted, found that Facebook users tended to get annoyed by viral content even when it was being shared by close friends.

Facebook likely wouldn't have to eliminate virality to deal with the problem, he continued, illustrating his thesis with a crude graph suggesting that the company could slash the bulk of viral harms without materially harming the platform's utility to users. The graph was hypothetical, Jin noted, because Facebook had never tried dialing back. Perhaps it was time.

Hegeman was skeptical. If Jin was right, he responded, Facebook should probably be taking drastic steps like shutting down all reshares, and the company wasn't in much of a mood to try. "If we remove a small percentage of reshares from people's inventory," Hegeman wrote, "they decide to come back to Facebook less."

Hegeman wasn't antagonistic toward Integrity staffers. Many of those who worked with him considered him an ally. But altering Facebook in ways that decreased its daily user count was simply unthinkable.

A few months after that exchange, the company revised a guide for testing product changes and how to interpret experiment results. Whatever evolution in the company's thinking had taken place in

the years since the 2016 election—whether it was greater attention to content quality or a better understanding of the perils of
virality—"Daily Average People" was still the company's North Star.
The guide said any proposed feature that reduced the number of
daily Facebook users by even 0.1 percent was almost certainly dead
on arrival.

"Rule #1, DAP is sacred," the presentation declared.

The company tried not to touch other metrics, like the number
of sessions logged, time spent, and original posts produced. But
when it came to misinformation metrics or the other harms that
Jin was looking at, there weren't yet guardrails that would prevent
growth-focused teams from changing Facebook in ways that damaged integrity metrics, or even warn when they did. It took until the
second half of 2019 for someone to build a dashboard that would
detect when other product teams were undercutting Integrity staffers' work.

As soon as the new system was live, Facebook's Integrity team
realized that it had a problem. Growth-focused changes to the platform appeared to be actively making integrity problems worse. A
memo by a Civic researcher noted that, over the course of the year,
content from pages that routinely posted misinformation was up
17 percent. Views of content from pages that had broken Facebook's
community standards at least four times over the past ninety days
were up 64 percent.

After two and a half years of self-examination, the idea that pro-
growth changes sometimes made integrity problems worse couldn't
be a complete surprise to anyone inside Facebook. But nobody had
been able to see how directly opposed the two goals were. Since at
least the end of 2018, and almost certainly long before that, Civic's
colleagues elsewhere in Facebook had been racking up usage gains
and earning bonuses by favoring content that the company had
publicly pledged to fight.

With the 2020 presidential primaries approaching, Civic argued
that the company finally needed to figure out how to stop shooting
its own integrity efforts in the foot.

"Crucial conversations are needed," the researcher wrote, to determine whether Facebook really believed that the benefits of an extra 1 percent growth outweighed the harm of showing users 10 percent more misinformation.

From Civic's point of view, the answer to that question was obvious. Despite the team's efforts, views of content from pages that had repeatedly violated Facebook's rules had risen by 23 percent since the beginning of the year. Heading into the 2020 election, the company could hardly afford the continued erosion of its defenses.

"There's a need to hold the line," the researcher wrote in a memo. Undoing the growth-focused product changes most responsible for that increase was as urgent as the need to "put up sandbags in a storm."

If Civic had thought Facebook's leadership would be rattled by the discovery that the company's growth efforts had been making Facebook's integrity problems worse, they were wrong. Not only was Zuckerberg hostile to future anti-growth work; he was beginning to wonder whether some of the company's past integrity efforts were misguided.

Empowered to veto not just new integrity proposals but work that had long ago been approved, the Public Policy team began declaring that some failed to meet the company's standards for "legitimacy." Sparing Sharing, the demotion of content pushed by hyperactive users—already dialed down by 80 percent at its adoption—was set to be dialed back completely. (It was ultimately spared but further watered down.)

"We cannot assume links shared by people who shared a lot are bad," a writeup of plans to undo the change said. (In practice, the effect of rolling back Sparing Sharing, even in its weakened form, was unambiguous. Views of "ideologically extreme content for users of all ideologies" would immediately rise by a double-digit percentage, with the bulk of the gains going to the far right.)

"Informed Sharing"—an initiative that had demoted content shared by people who hadn't clicked on the posts in question, and which had proved successful in diminishing the spread of fake news—was also slated for decommissioning.

"Being less likely to share content after reading it is not a good indicator of integrity," stated a document justifying the planned discontinuation.

A company spokeswoman denied numerous Integrity staffers' contention that the Public Policy team had the ability to veto or roll back integrity changes, saying that Kaplan's team was just one voice among many internally. But, regardless of who was calling the shots, the company's trajectory was clear. Facebook wasn't just slow-walking integrity work anymore. It was actively planning to undo large chunks of it.

Jin's calls got more urgent—and more basic.

A few months after his first warning that the company's pursuit of viral growth was harming user safety and content quality, he wrote another memo, this one titled "Defining Success in Addressing Integrity Harms."

The engineer began with a simple premise: Facebook would never be able to remove every bad thing from its platform, and it was unreasonable to expect it to. But the company obviously had some responsibility—so what did it mean to succeed at combating integrity problems?

"Success is addressing all places where our products are disproportionately amplifying harms relative to a world in which they did not exist," he wrote.

The principle was one that children tend to learn early. The previous cleanliness of the kitchen floor isn't relevant to whether you spilled something on it. Likewise, the story of how pouring yourself a beverage ended with an overturned jug might be complicated, even worthy of rigorous study, but these questions aren't relevant to whether you are the party who needs to fetch a mop.

For Facebook, Jin wrote, cleaning up the messes it made meant that it needed to pay particular attention to its recommendation systems and features that encouraged bad behavior or were disproportionately prone to abuse. To avoid a disaster in the 2020 elections, he wrote, the company would either have to cut back on the features that amplified social problems or get better at plucking out the bad stuff.

Facebook preferred the latter approach, Jin noted, but it was technically much harder to pull off. Besides, cleaning up Facebook and Instagram by targeting misbehavior raised inevitable concerns about censorship. No matter how much money Facebook spent on the effort, it still risked losing control of its platforms.

In contrast, Jin continued, Facebook could be certain of meeting its goals for the 2020 election if it was willing to slow down viral features. This could include imposing limits on reshares, message forwarding, and aggressive algorithmic amplification—the kind of steps that the Integrity teams throughout Facebook had been pushing to adopt for more than a year. The moves would be simple and cheap. Best of all, the methods had been tested and guaranteed success in combating longstanding problems.

The correct choice was obvious, Jin suggested, but Facebook seemed strangely unwilling to take it. It would mean slowing down the platform's growth, the one tenet that was inviolable.

"Today the bar to ship a pro-Integrity win (that may be negative to engagement) often is higher than the bar to ship pro-engagement win (that may be negative to Integrity)," Jin lamented. If the situation didn't change, he warned, it risked a 2020 election disaster from "rampant harmful virality."

As pessimistic as Jin's appraisal of Facebook's elections preparation was, it was still worded delicately. Among integrity staffers, it was commonplace to observe that Facebook's classifiers were no match for the granular enforcement tasks they were being asked to perform.

Even including downranking, "we estimate that we may action as little as 3–5% of hate and 0.6% of [violence and incitement] on Facebook, despite being the best in the world at it," one presentation noted. Jin knew these stats, according to people who worked with him, but was too polite to emphasize them.

Civic and old News Feed hands like Jin weren't the only ones getting worried by the fall of 2019. Plenty of researchers who got nowhere near questions of platform mechanics were getting spooked. Though QAnon—the theory that, even though President Trump was spending a record-setting number of days golfing, he was also locked in mortal combat with the malevolent, pedophilic forces of the Deep State—originated on 4Chan and was incubated on Reddit, Facebook groups had provided it with the hothouse conditions for exponential growth. Some of that movement's growth came from the hyperactivity of its adherents. Those who bought into the conspiracy theory's framing of Trump's battle against evil spread it with messianic fervor.

Some of it, too, came from Facebook's algorithms. Just as Facebook had learned that its "Groups You Should Join" was fueling German extremism in 2016, the platform appeared unusually fond of recommending QAnon.

The phenomenon wasn't a surprise: recommendation systems like Facebook's inevitably try to pigeonhole users, and the more obscure their interests, the better. Knowing someone follows Greco-Roman wrestling helps Facebook personalize their feed more than knowing they like college football. Conspiracy theories such as QAnon have exactly this kind of narrow-but-deep appeal.

Company researchers used multiple methods to demonstrate QAnon's gravitational pull, but the simplest and most visceral proof came from setting up a test account and seeing where Facebook's algorithms took it.

After setting up a dummy account for "Carol"—a hypothetical forty-one-year-old conservative woman in Wilmington, North Carolina, whose interests included the Trump family, Fox News, Christianity, and parenting—the researcher watched as Facebook guided Carol from those mainstream interests toward darker places.

Within a day, Facebook's recommendations had "devolved toward polarizing content." Within a week, Facebook was pushing a "barrage of extreme, conspiratorial, and graphic content."

The problem wasn't just News Feed. It was page recommendations, suggestions to look at "similar posts," and even notifications. The researcher's write-up included a plea for action: if Facebook was going to push content this hard, the company needed to get a lot more discriminating about what it pushed.

Later write-ups would acknowledge that such warnings went unheeded.

Of the millions of Facebook users who would eventually join groups devoted to the conspiracy, it's a safe bet that many weren't actual adherents. But by the fall of 2019, QAnon true believers had already committed a few weird crimes—the blockading of a bridge across the Hoover Dam, a kidnapping, the occupation of a cement plant—and the company's user experience researchers had more insight into the bizarre conspiracy than most.

One Facebook director recalled an encounter with a user in Las Vegas, a common destination for user researchers trying to get out of the Bay Area's cultural bubble to recruit subjects. The setup was unremarkable: one of the director's employees was interviewing a succession of Facebook users about their activity on the platform while the director and a colleague sat behind a one-way mirror in a rented market research office space. The subjects were intentionally diverse, and the professionally trained Facebook employee leading the interview kept her questions rigorously neutral. But one interview subject, a middle-aged white woman, didn't see things that way.

Her posts, mostly reshares, had been heavy on "Make America Great Again" and QAnon content, and she seemed suspicious about even the blandest questions about how she used Facebook.

The woman began gesticulating, then raised her voice. The director initially found the subject's intransigence funny. But as the interviewer tried and failed to calm the woman, he and his colleague began to exchange worried looks.

Holy shit, the director thought. *She's going to come across the table.*

Behind the one-way glass, the director and his colleague rose from their seats. They stayed standing as the interviewer hurriedly wrapped up, ready to burst into the room if the woman lunged at her.

"We internalized that," he said. "Our users are crazy."

With the 2020 election closing in and much of its work bogged down by the Policy and Product teams, Civic needed to do something bold. So it made a slide deck.

Embedded in a PowerPoint presentation that Samidh Chakrabarti, Kaushik Iyer, and others bounced around toward the end of summer, the color-coded chart matched the severity of potential crises with Facebook's level of readiness to address them. As the document began coming together, Chakrabarti shared it with Katie Harbath.

The presentation was grim, Harbath told him, but it wasn't wrong.

"I kind of want them to be scared," she recalled Chakrabarti telling her. Then he emailed the presentation directly to Zuckerberg and a collection of two dozen other executives who attended a weekly meeting in the CEO's glass-walled conference room, nicknamed the Aquarium.

White was for things that were not that important and which Facebook was adequately prepared for. Yellow and orange were for middle grounds, places where things could go wrong but the capacity for a true disaster was limited. Red was for high-profile problems on which Facebook was prepared to do only damage control.

There was one last category: crimson. There was a lot of it, even though it was reserved for problems that were both "very likely" to occur and severe enough to cause "social conflict," a phrase often used as a euphemism for violence. According to the chart, the company was completely unprepared for domestic manipulation attempts, fact-checking was too slow and limited to keep up with the speed at which falsehoods traveled, and the "perverse incentives" of Facebook's recommendation system were actively degrading the quality of news that people consumed. The company's efforts to prevent misconduct by large-scale pages and groups were woefully inadequate, and there was no way for users to tell when such entities were being run from overseas. Facebook's enforcement system was so unreliable that it could conceivably take wrongful action against one of the major-party presidential candidates.

Videoconferencing in from DC, Harbath watched Chakrabarti present the material to the assembled executive team, keeping her eyes mostly on Zuckerberg. The CEO didn't say much, Harbath recalled. He just looked frustrated. Before Chakrabarti had reached the end, Zuckerberg cut him off with a question: What did Chakrabarti need to address the identified problems?

Chakrabarti had a ready answer. He asked Zuckerberg to impose a "lockdown" until things were fixed.

Zuckerberg looked quizzical. Lockdowns were generally imposed from above, not requested from below. When the company declared one for a particular crisis, anyone working on the problem was required to work nights and weekends until they achieved specific "exit criteria," meaning they had gotten the problem under control. The process was supposed to be miserable, but it came with one advantage. When a team was in lockdown, it had the ability to dragoon staff and resources until the problem at hand was behind the company.

If Chakrabarti thought that a lockdown was necessary, the company would do it, Zuckerberg said. Then he adjourned the meeting.

As executives filed out, Zuckerberg pulled Integrity's Guy Rosen aside. "Why did you show me this in front of so many people?" Zuckerberg asked Rosen, who as Chakrabarti's boss bore responsibility for his subordinate's presentation landing on that day's agenda.

Zuckerberg had good reason to be unhappy that so many executives had watched him being told in plain terms that the forthcoming election was shaping up to be a disaster. In the course of investigating Cambridge Analytica, regulators around the world had already subpoenaed thousands of pages of documents from the company and had pushed for Zuckerberg's personal communications going back for the better part of the decade. Facebook had paid $5 billion to the U.S. Federal Trade Commission to settle one of the most prominent inquiries, but the threat of subpoenas and depositions wasn't going away.

Should the 2020 election sour on Facebook's watch, regulators would want to know whether Zuckerberg was aware of problems ahead of time, and they would likely be asking the question under oath. The last thing Facebook lawyers wanted was evidence that the CEO had been told it was a mess and had shrugged it off. Chakrabarti had just created a paper trail to that effect, and there was no shortage of witnesses. A company spokesperson denied that Zuckerberg's displeasure had anything to do with being shown an alarming document in front of a significant internal audience. Instead, the company said, he believed that the material hadn't been adequately vetted.

If there had been any doubt that Civic was the Integrity division's problem child, lobbing such a damning document straight onto Zuckerberg's desk settled it. As Chakrabarti later informed his deputies, Rosen told him that Civic would henceforth be required to run such material through other executives first—strictly for organizational reasons, of course.

Chakrabarti didn't take the reining in well. A few months later, he wrote a scathing appraisal of Rosen's leadership as part of the company's semiannual performance review. Facebook's top integrity official was, he wrote, "prioritizing PR risk over social harm."

Harbath also lost standing. She had been one of Public Policy's main liaisons with Civic, and Rosen had already told other executives he had lost faith in her, seeing her as a holdover from a time when the company's response to anything involving politics was a hearty "yes" and extensive public discourse. Elections were no longer an opportunity, but a problem to be managed—discreetly, if possible.

To keep Facebook's elections and integrity work under control, Facebook turned to Molly Cutler, the company's head of Strategic Response. Cutler had risen to prominence as an assistant general counsel drafted to steer the company through the Cambridge Analytica mess. When that died down, she had been made a sort of permanent crisis manager. Unlike Rosen, she was more explicit about why the company was creating a new layer of management separating election-related work from the C-suite, according to two executives. The purpose was to shield Zuckerberg. (The company attributed the change to a need to ensure that executives had access to "clear, accurate, and unbiased information.")

The strictures of the new regime readily became apparent. Soon after Cutler assumed her new role, Zuckerberg asked a question about Facebook's ad library on a group email thread. The question wasn't anything sensitive, and Harbath, who had been involved in setting the thing up and explaining it to the public, answered it.

"Molly tore me to shreds," Harbath said. "It was, 'You do not talk to Mark without talking to me.'"

Soon after, Harbath was relieved of her role running Facebook's Global Elections team and reassigned to "Escalations," the division responsible for trying to quickly resolve content moderation issues prominent enough to generate bad press or anger powerful people. She had been relegated, in effect, to high-level customer service.

In a reflection of the skill set the company was looking for to fill her old role, Facebook replaced Harbath with Heather King, a disaster response specialist who had worked for the National Security Council and the Federal Emergency Management Agency. (King never spoke in public during her time at Facebook. She went to the Department of Defense after the 2020 election, where her official biography mentions only that she worked "in the technology sector.")

Regardless of the fallout, Chakrabarti got his lockdown. Over the next three months, Civic and its dragooned helpers achieved more than the team had in the first nine months of the year. Instagram established, for the first time, a process for removing inauthentic accounts. Facebook overhauled incentives for its fact-checking program to encourage its partners to tackle viral misinformation instead of, say, inspirational quotes misattributed to Winston Churchill.

The platform started building tools to identify "narrowcast" propaganda efforts, in which manipulation efforts targeted a specific minority community. It began enforcing anti-spam policies in Facebook groups and banned ads that encouraged people not to vote.

On the technical side, the company had at last built systems that would allow it to detect a mass-scale voter suppression effort and respond to it.

When the lockdown lifted, there was progress: almost all the dark red had faded to a shade signifying the company was at least aware of the risks it faced. But the updated version of the chart that Chakrabarti had lobbed in front of Zuckerberg didn't look as different as one might have hoped. Some of the shortfalls were technical—Instagram still hadn't gotten its integrity work fully up to speed—but others reflected either indecision by company leadership or the affirmative choice not to do work that Civic considered essential.

Facebook still hadn't given Civic the green light to resume the fight against domestically coordinated political manipulation efforts. Its fact-checking program was too slow to effectively shut down the spread of misinformation during a crisis. And the company still hadn't addressed the "perverse incentives" resulting from News Feed's tendency to favor divisive posts. "Remains unclear if we have a societal responsibility to reduce exposure to this type of content," an updated presentation from Civic tartly stated.

"Samidh was trying to push Mark into making those decisions, but he didn't take the bait," Harbath recalled.

There was one final election risk that was hard to predict: Donald Trump. Though his name did not appear in the elections preparation assessment, nobody was confused about what Civic meant when it cited the "emerging risk" of "aggressive techniques by domestic campaigns." The company would need to demonstrate that it was serious about enforcing its rules against domestic figures, the presentation advised, and "respond quickly and consistently" when its rules were broken.

This plan was more than a little aspirational. While the company's community standards didn't grant politicians any extra slack, Facebook had largely shied away from enforcing its rules against any but the most fringe candidates and officeholders. Most of this was tacit,

but in late September 2019, Facebook's head of Global Affairs—
former British deputy prime minister Nick Clegg—gave a talk noting that the company had decided to exempt both politicians' posts
and campaign ads from Facebook's fact-checking program.

The announcement kicked off a multiday media frenzy, but
reporters weren't the only party worked up by the prospect of Facebook running false advertisements. Though the issue hadn't been
a problem in the past—Trump's false statements didn't need paid
media to dominate social media and cable TV news—Chakrabarti
and his team viewed Facebook's decision as an abdication of basic
responsibility. Surely there had to be some line that even a politician
couldn't cross.

Chakrabarti had been strenuously vocal in his opposition before
Clegg's announcement, which was why Harbath got nervous when
she heard he was planning to address it again during Civic's weekly
all-hands election meeting. That week Harbath, who normally listened in, was joined by far more of her Public Policy colleagues than
normal.

Chakrabarti began by extolling Facebook's mission of giving
people a voice and explained how he had joined the company to
further it. Disagreements on how to do that were to be expected, he
said, but the decision to run political ads with no oversight wasn't
defensible. Taking money for spreading falsehoods, he said, was a
betrayal of the platform's users.

As he walked through his combative prewritten remarks, Civic
staffers looked between each other to confirm what seemed increasingly obvious: this was a resignation speech, and Chakrabarti was
torching Facebook's leadership on his way out the door.

Harbath had the same thought, and she wasn't happy about it.
Her relationship with Chakrabarti had grown strained as Public
Policy and Civic had withdrawn into warring camps, and she had
started to suspect he might be withholding information from her
about Civic's work. But she was the closest thing to an ally Civic had
on the Washington Policy team—quitting without telling her was
messed up.

She needn't have worried. After lambasting his employer's lack of judgment and spine, Chakrabarti made a pivot. Civic's work was too important to quit, he said, and he hoped his staff would keep fighting alongside him. The audience—at least the portion of it that worked for Civic—responded with heavy applause.

Afterward, Chakrabarti told Harbath he had made the speech because he was worried that Civic's recurring defeats were breeding a sense of dejection among his staff. Getting through the 2020 election season would be miserable—almost no one on Civic thought Facebook was preparing adequately—and the loyalty to their paychecks wasn't enough. Civic might as well embrace its role as the company's loyal opposition, Chakrabarti said, devoted to mission above management.

The whole plan hinged on one thing: Facebook's leadership believing that Civic was indispensable. For the time being, at least, that was right. According to Harbath, Cutler remarked that she would have pushed for Chakrabarti's ouster if she didn't expect a substantial portion of his team would mutiny. (The company denies Cutler said this.)

10

Despite Facebook's public travails, business was booming. In 2016, it had booked $27 billion in revenue and $10 billion in profit. Both those numbers doubled over the course of the next two years, and, despite spending billions on content moderation and regulatory fines in 2019, the company's margins were still among the highest of the S&P 500. Usage of the company's products grew every quarter.

Outside the view of reporters and financial analysts, however, internal benchmarks of the company's health were beginning to falter. Users were posting original content a third less often than they had five years prior, a gap that Facebook sought to fix by increasing the amount of reshared content it showed people and then by copying Snapchat's ephemeral messaging feature, Stories.

Those updates worked, for a while. But Facebook's namesake platform had deeper problems, especially among the young. In 2016, internal research showed that, at a time when overall use of Facebook Messenger was up 40 percent worldwide, the number of messages sent by young users in major Western markets was dropping. Those young users were connecting with fewer people, too, adding 36 percent fewer friends compared with the year before.

For a while, the company had hoped that younger teens were just taking their time getting on board—an analysis of teen user growth was optimistically illustrated with a sloth crawling across a road. By the time of a 2018 report on the "State of Teens," however, the numbers were clear. In 2012, there were 0.8 accounts on the platform for every fourteen-year-old. That number had since been cut in half.

Among young teens, Facebook users were a minority. Kids were the canaries in Facebook's attention mine, the report warned, the first to disappear "when products become stale."

Falling youth activity on Facebook's platform was a potentially dire long-term problem for the company, though the report noted a saving grace. The platform's most formidable competitor was Instagram, which Facebook conveniently owned. As long as Instagram could keep young users engaged with the company's products, its sister platform's trouble with teens looked manageable.

Along with its rosier growth outlook, Instagram also enjoyed a far better public image than Facebook as a whole. Nobody had accused the photo-based platform of throwing elections, betraying user privacy, or fueling genocide. That founders Mike Krieger and Kevin Systrom remained at the app's helm until late 2018 didn't hurt. Most Instagram users didn't even know the app was owned by Facebook.

There was another reason why Instagram had avoided many of Facebook's travails. Its far simpler mechanics simply couldn't go wrong in many ways that Facebook's had. Fake news publishers were far less threatening on Instagram, because the platform didn't allow users to insert links into their posts. The lack of pages and groups forestalled the need to defend against complex cross-product manipulation attempts. And all of Facebook's struggles with reshared content weren't a concern on an app entirely lacking a reshare button.

Though Instagram's design rendered it immune to much of the criticism lobbed at Facebook, the platform's younger user base and image-focused culture made it more vulnerable than Facebook on one front: mental health.

Fears that new forms of media will corrupt youth are perennial, so it was perhaps inevitable that smartphones and social networks would trigger the same sort of hand-wringing that video games, television, and comic books had generations before. But the interactivity and competitiveness of social media added a new twist. By 2018, the *Journal of Family Medicine and Primary Care* had cataloged 259 selfie-related deaths, most the result of what researchers termed "risky behavior."

Plenty of rival social media apps drew heat when their users got too comfortable near a cliff or a herd of buffalo. But cultural critics and clinicians had also begun looking at whether social media might be contributing to ills such as bullying, eating disorders, and unrealistic views of others' lives. Instagram, by virtue of its vast user base and emphasis on self-presentation, took the worst of it. By 2017, a British study had found that Instagram had the worst effect of any social media app on the health and well-being of teens and young adults.

Facebook needed to look into the matter itself. By late 2017, Instagram had created a Well-Being team with a mandate to examine possible harms to users and how the platform could mitigate them.

The team's work soon received a boost from two vastly different events. The first was Adam Mosseri's appointment as head of Instagram in October 2018 following the departure of the app's founders. Whether because he was a parent, enthusiastic about youth culture, or simply concerned about the welfare of teens, Mosseri personally was a staunch backer of exploratory work, giving the Well-Being team unusual stature within Instagram's organization.

The second was the death of Molly Russell, a fourteen-year-old from North London. Though "apparently flourishing," as a later coroner's inquest found, Russell had died by suicide in late 2017. Her death was treated as an inexplicable local tragedy until the BBC ran a report on social media activity in 2019. Russell had followed a large group of accounts that romanticized depression, self-harm, and suicide, and she had engaged with more than 2,100 macabre posts, mostly on Instagram. Her final login had come at 12:45 the morning she died.

"I have no doubt that Instagram helped kill my daughter," her father told the BBC.

Later research—both inside and outside Instagram—would demonstrate that a class of commercially motivated accounts had seized on depression-related content for the same reason that others focused on car crashes or fighting: the stuff pulled high engagement. But serving pro-suicide content to a vulnerable kid was clearly indefensible, and the platform pledged to remove and restrict the recommendation of such material, along with hiding hashtags like #Selfharm. Beyond exposing an operational failure, the extensive coverage of Russell's death associated Instagram with rising concerns about teen mental health.

Instagram was uncertain those concerns were valid, so the platform's Well-Being staff moved promptly into action to find out. By the second half of 2019, they had put together a slew of internal research, a mix of large-scale user surveys, qualitative efforts to understand how teens experienced the platform, and analyses of their consumption habits. For good measure, Instagram also commissioned research from the polling firm YouGov.

Some of the findings were reassuring if unsurprising. Teens reported that swapping messages with friends, looking at comedy accounts, and learning about things on Instagram was good for them. When seeking a distraction, teens overwhelmingly considered the app to be beneficial.

There were other areas where the platform seemed much less constructive. Though much attention, both inside and outside the company, had been paid to bullying, the most serious risks weren't the result of people mistreating each other. Instead, the researchers wrote, harm arose when a user's existing insecurities combined with Instagram's mechanics. "Those who are dissatisfied with their lives are more negatively affected by the app," one presentation noted, with the effects most pronounced among girls unhappy with their bodies and social standing.

There was a logic here, one that teens themselves described to researchers. Instagram's stream of content was a "highlight reel," at once real life and unachievable. This was manageable for users who arrived in a good frame of mind, but it could be poisonous for those who showed up vulnerable. Seeing comments about how great an acquaintance looked in a photo would make a user who was unhappy about her weight feel bad—but it didn't make her stop scrolling.

"They often feel 'addicted' and know that what they're seeing is bad for their mental health but feel unable to stop themselves," the "Teen Mental Health Deep Dive" presentation noted. Field research in the U.S. and U.K. found that more than 40 percent of Instagram users who felt "unattractive" traced that feeling to Instagram. Among American teens who said they had thought about dying by suicide in the past month, 6 percent said the feeling originated on the platform. In the U.K., the number was double that.

"Teens who struggle with mental health say Instagram makes it worse," the presentation stated. "Young people know this, but they don't adopt different patterns."

These findings weren't dispositive, but they were unpleasant, in no small part because they made sense. Teens said—and researchers appeared to accept—that certain features of Instagram could aggravate mental health issues in ways beyond its social media peers. Snapchat had a focus on silly filters and communication with friends, while TikTok was devoted to performance. Instagram, though? It revolved around bodies and lifestyle. The company disowned these findings after they were made public, calling the researchers' apparent conclusion that Instagram could harm users with preexisting insecurities unreliable. The company would dispute allegations that it had buried negative research findings as "plain false."

Concern that Instagram had become a peer pressure cooker wasn't restricted to do-good social scientists. Product teams focused on the health of Instagram's business had cause for concern, too. Social competition had long been a part of the Instagram experience, but both user interviews and behavioral data seemed to suggest that many users felt the bar for a good post had become unattainably high. Affluent white users in the United States in particular had evidently responded by cutting back on posting. They increasingly favored "Stories," which disappeared after a day—material that, from a product standpoint, had a shorter shelf life.

Teens were worried about the "myth" that they had to look perfect in images, researchers focused on increasing usage found. They were still consuming plenty of content, of course, but Instagram needed them producing it, too. Trying to relieve anxiety around posting was a business imperative.

Figuring out how to alleviate negative social comparison and body image concerns was also simply the right thing to do, of course. Mosseri wanted to identify the specific features that might be contributing to the problem.

One suspect: the heart-shaped like button.

Like its parent company, Instagram had implemented algorithmic content ranking, though later than Facebook, in 2016. That meant that, even setting aside celebrity and professional influencer content, a user was almost guaranteed to see content from friends and acquaintances that was more engaging than whatever they themselves were posting.

"People see about 5% as many likes on their own posts as those they see on IG," company research found. Like counts quantified a popularity gap, and users reported deleting pictures of themselves that failed to impress their peers and Instagram's algorithms. Mosseri announced in November 2019 that the platform would begin experimenting with hiding likes.

Some influencers hated the idea. Like counts were proof to their sponsors and fans that they were popular. But according to people who worked on the effort, known as "Project Daisy," the company paid little attention and was uncharacteristically prepared to accept an engagement loss if it succeeded. The benefits of potentially making users feel better and post more content were too exciting. As Mosseri put it when announcing the experiment, "We will make decisions that hurt the business if they help people's well-being and health."

Even at the time, the statement was eye-catchingly bold—all the more so because Instagram's findings were earlier and less concrete than the company's conclusions regarding the flaws of its recommender systems and Facebook's vulnerability to large-scale manipulation. But when it came to mental health, the company wasn't vaguely talking about "tradeoffs." Mosseri was saying that Instagram might have inadvertently been built with a feature that helped its business but harmed users—and promising that, if it had, the company would fix it.

On October 17, 2019, Zuckerberg took to the ornate podium at Georgetown's Gaston Hall to deliver a rare public address, on the subject of free speech. The grand hall's murals and Jesuit crests gave the event the feel of a sermon, and Facebook's PR team had certainly pitched it as such.

Zuckerberg began by acknowledging the death earlier that day of Maryland congressman Elijah Cummings, whose illustrious civil rights career began at age eleven, when he was attacked by a white mob while integrating a Baltimore pool, leaving him with a lifelong facial scar.

"He was a powerful voice for equality, social progress, and bringing people together," Zuckerberg told the crowd. Then he made the case that Facebook stood for those values, too.

Zuckerberg said he had created Facebook because he believed that progress came from regular people having a voice, crediting that conviction to being in college at the time of the Iraq War. "I remember feeling that if more people had a voice to share their experiences, maybe things would have gone differently," he said. "Those early years shaped my belief that giving everyone a voice empowers the powerless and pushes society to be better over time."

But, Zuckerberg said, he was worried that the commitment to free, democratic speech in America might be wavering. "In times of social turmoil, our impulse is often to pull back on free expression," Zuckerberg told the crowd. "We want the progress that comes from free expression, but not the tension."

Amid an onslaught from legislators, the media, and the public, Zuckerberg was doubling down on his, and his platform's, preference for free speech. He made clear he was no absolutist, citing Facebook's commitment to curbing terrorist propaganda, the bullying of young people, and pornography. But, beyond that, he asked, "Where do you draw the line?

"Most people agree with the principles that you should be able to say things other people don't like, but you shouldn't be able to say things that put people in danger," Zuckerberg continued, before making a long argument that expanding the definition of "dangerous speech" could be risky.

Facebook had built systems, many of them powered by AI, to address around twenty categories of harmful content, he said. "All of this work is about enforcing our existing policies, not broadening our definition of what is dangerous," he said.

When it came to misinformation, rather than directly addressing falsehoods on the platform, Zuckerberg said, the company had found a better strategy: making sure the accounts were authentic, and removing those that were not, including bots. The true fight wasn't against polarization and misinformation, he said. It was against those who "no longer trust their fellow citizens with the power to communicate and decide what to believe for themselves."

And there, Zuckerberg said, Facebook would hold the line. Social media was "a Fifth Estate," he declared, giving its users the ability to speak up against the "traditional gatekeepers in politics or media." A social network was a uniquely democratic force, incompatible with a repressive government like that of China, and Americans needed to stand up for it. "Democracy depends on the idea that we hold each other's right to express ourselves and be heard above our own desire to always get the outcomes we want," Zuckerberg said. "We need to make sure we're empowering people, not simply reinforcing existing institutions and power structures."

Social media would, left to its own devices, bring about a new era of freedom and social progress, he concluded. "I believe in giving people a voice because, at the end of the day, I believe in people," Zuckerberg said. "From all of our individual voices and perspectives, we can bring the world closer together."

If Zuckerberg's rendition of Facebook's anti-war origin story was revisionist, the CEO's description of Facebook's current purpose was no more accurate. As much as the company talked about giving people a voice, the company had been built to make people use Facebook—and then repeatedly refined to ensure they used it more.

Equally nonsensical was Zuckerberg's claim that the company could control misinformation and manipulation attempts by removing fake accounts. By late 2019, the idea that the platform's problems could be pinned on bots had been thoroughly discredited inside the company and out. Whatever one might think of the people fanatically promoting the QAnon conspiracy or the claim that vaccines cause autism, they were real people. Facebook just allowed zealots to earn clout on their platform far in excess of what they could achieve anywhere else.

For people working on platform integrity issues, the speech was clarifying. Zuckerberg, they realized, personally did not support their work. He had once declared himself the only guy capable of fixing Facebook. Now he was publicly concluding that it no longer needed fixing.

Arturo Bejar, on his second stint at Facebook, working as a consultant at Instagram, watched the speech with some dismay. "My experience of working with him," he said of Zuckerberg, "is that he's very good at working with principles like free speech. But I never got the sense that he understood that what he was really dealing with, and was responsible for, was humans.

"This is a banner thing for Mark, and a banner for the people close to him," he continued. "There's this sense that if we let people say anything, everything's going to be okay." Bejar had once believed that himself. But now he had come to see that Facebook bore responsibility for people's experiences on its platform.

The Georgetown students clapped respectfully once Zuckerberg was done. Over on Facebook, where a far larger audience had been watching, a markedly different reaction played out. People were doing the digital equivalent of a standing ovation, hooting and hollering their support. Zuckerberg was a genius, a humanitarian, and a world leader, declared users in hundreds of short and oddly similar sentences featuring the words "thanks," "love," or "congratulations." Out of hundreds of user responses, only a half dozen were hostile.

"Thank you for creating the Best social platform ever love youuuuuuuuu 😘😘😘😘😘," wrote one user. "I LOVE BIG BOY VERY SMART MARK," wrote another.

Were the fawning comments faked? Written by an army of foreign bots? Censored by Zuckerberg's staff in near real time? Facebook's Communications department denied anything was amiss, but a web-scraping tool revealed that the sentiments being shown on screen weren't representative of the 45,000 users who had commented.

A closer look at the handful of negative comments that Facebook users did see offered a clue as to what was really happening. The only nastiness was expressed as sarcasm, people "thanking" Zuckerberg for ruining the country and suppressing their speech.

Facebook's communications team declined to explain what it had done beyond saying that the company tried to preference "quality" responses. But it seemed clear what was happening. Facebook had deployed a comment-filtering system to prevent the heckling of public figures such as Zuckerberg during livestreams, burying not just curse words and complaints but also substantive discussion of any kind. The system had been tuned for sycophancy, and poorly at that.

The irony of heavily censoring comments on a speech *about* free speech wasn't hard to miss. But in the context of other allegations about Facebook—a company that stood accused of subverting elections, brainwashing users, and fueling genocide—this hardly had the makings of a front-page story. It wasn't a surprise, then, that the *Wall Street Journal*'s editors didn't leap at the opportunity to probe whether the company had crudely silenced voices criticizing "Mark Suckerberg."

That hadn't stopped me from trying. The truth was that my story pitches had been flopping with some regularity. I was eight months into arguably one of the marquee beats in tech journalism, and I began to wonder if it was time to move on from covering Facebook.

My colleagues in San Francisco covered companies helmed by men who announced fictional buyouts at $420 a share (Tesla's Elon Musk) or launched bottles of tequila through glass doors to unwind (WeWork's Adam Neumann). In contrast, paparazzi caught Mark Zuckerberg riding an electric surfboard while wearing too much sunscreen. Other staples of *Wall Street Journal* reporting were similarly hopeless when it came to Facebook. The company was already too big to consider major acquisitions, Zuckerberg's age precluded stories about successor jockeying, and the board was pliant. I'd learned after a few wild goose chases that rumors of Sandberg's imminent departure flew with the same predictability as clay pigeons at a shotgun range. Bereft of executive drama, I might have eked stories out of Facebook's financial statements, but the business was both uncommonly profitable and stable, bringing in hundreds of millions of dollars a day, one click at a time.

Still, something didn't sit right with me about the episode. Admittedly, the internal mechanics of a social media platform—which comments get surfaced and which don't, why some posts go viral and some go unnoticed, how memes are perpetuated—none of this was particularly sexy stuff. It was bogged down in the language of "algorithms," "classifiers," and "user experience"—terms all but designed to put the uninitiated to sleep. Moreover, they were concepts I myself would have to better grasp if I were to share them with readers.

But the plain fact was that those mechanics were central to what Facebook had become as a product, which was a machine that served you content custom-designed to keep you scrolling, ad infinitum. We reporters had devoted countless column inches to the hate speech, fake news, and revenge porn that had by then entered the popular consciousness. We had covered controversial Facebook rules and content moderation calls. But, as I came to see, which posts were removed was far less consequential than which were amplified.

A petty obsession with the rigged comments section of Zuckerberg's speech became a kind of North Star. Beneath the surface of a platform where users could post and interact was a hell of a lot of machinery. Servicing it would surely require professional mechanics, people who understood how the components worked and what to do when they broke. I didn't know anybody with that skill set—not yet—but at least I knew who I was looking for.

—

An object lesson in the perils of amplification took place just a few months after Zuckerberg's speech and my minor epiphany. Brandon Silverman was walking down a hall in Menlo Park when he ran into an executive who worked closely with Facebook's C-suite.

"You kicked off quite a fire drill yesterday," Silverman's colleague told him, with an expression somewhere between a smirk and a grimace. Silverman was confused. He hadn't been in touch with anyone in the company's senior ranks. What had he done?

"Tuesday's email," the man said. Then Silverman remembered.

A few months before, he had set up an automated email in CrowdTangle to send a daily list of the platform's ten most widely viewed posts to a handful of senior executives as part of his effort to focus leadership attention on content quality. Then he'd forgotten all about it.

CrowdTangle's rundown of that Tuesday's top content had, it turned out, included a butthole. This wasn't a borderline picture of someone's ass. It was an unmistakable, up-close image of an anus. It hadn't just gone big on Facebook—it had gone *biggest.* Holding the number one slot, it was the lead item that executives had seen when they opened Silverman's email. "I hadn't put Mark or Sheryl on it, but I basically put everyone else on there," Silverman said.

The picture was a thumbnail outtake from a porn video that had escaped Facebook's automated filters. Such errors were to be expected, but was Facebook's familiarity with its platform so poor that it wouldn't notice when its systems started spreading that content to millions of people?

Yes, it unquestionably was. The massive gap in Facebook's self-awareness was why Silverman had set up the alert in the first place. Winning content was often terrible, spammy, and in violation of platform rules. That day's email had just been an especially graphic illustration of the problem.

The incident spurred some internal discussion at Facebook, and Silverman was called in to help come up with a functional definition of content quality. Defining what "quality" meant to Facebook was a first step in getting to the heart of a matter it had never really explored. Confronting what it didn't want on its platform was one thing; thinking about what it actually *wanted* there was another. While Facebook was comfortable removing or downranking bad content, the company had never considered upranking good content.

Silverman considered the distinction semantic. Whatever content Facebook served at the top of users' feeds had been functionally upranked by the News Feed algorithm, no matter what the company called it. The whole process was artificial, so why not promote things that Facebook might be proud of?

"They didn't get to what I was hoping to get to, which was planting the flag on what Facebook thinks the feed should be made up of," Silverman said. "Mark wanted the flexibility."

The episode didn't draw attention outside Facebook's walls. The tumult that would ensue in the coming months and years would be much harder to miss.

11

When COVID-19 hit in early 2020, Facebook was one of the first major U.S. companies to shut down. While skeleton crews stayed on site at company data centers to keep the company's servers running, Facebook sent the overwhelming majority of its 50,000 employees home to await orders and come to grips with the reality of a global pandemic.

It was a hard time for people in general, and for many employees personally. But for the company? The pandemic was nothing short of a boon. After three years mired in scandals over fake accounts, moderation flubs, and misinformation, Facebook as a product suddenly felt indispensable for millions of people sheltering at home, seeking any sort of contact with other humans, aside from physical. When Italy became the first country after China to impose a nationwide COVID lockdown in March, use of Facebook products rose by 70 percent, with some, such as WhatsApp group calls, spiking tenfold. For businesses, too, the company's products were a lifeline. Facebook had been trying for years to push more commerce onto its platform, and the pandemic effectively did it.

Internal dashboards showed a usage windfall in the early days of the COVID lockdown. U.S. Facebook users alone increased their posting by half and spent 200 million hours on the app each day.

It would take time for the attention bonanza to pay off in ad revenue, but with $60 billion in cash and securities on hand, the company had plenty of money to throw around. It paid its outsourced content moderation staff to sit at home idle for months, and granted a $1,000 bonus to full-time staff. It also suspended performance reviews and said it would grade all employees as "exceeds expectations," guaranteeing everyone a hefty annual bonus.

The pandemic also ended a hiring drought. Facebook had been struggling to find enough employees to maintain its rapid growth, and Zuckerberg had "made it very clear" to senior Recruiting executives at the end of 2019 that he was upset by the pace of hiring, a later HR memo noted. "We knew we had a lot of products to build, and we had more confidence than other companies that we were going to be able to navigate the COVID-19 business uncertainty," it read. "And the bet paid off."

The world was changing in a way that just happened to suit Facebook's business model—and Zuckerberg's personal strengths.

The CEO had long been interested in pandemics. He and his wife, Priscilla, had launched the Chan Zuckerberg Biohub in 2016, with a mission to "support the science and technology that will make it possible to cure, prevent, or manage all disease by the end of the century." Zuckerberg was particularly interested in immunization, as it involved technology and, above all, scale. To run the Biohub, Zuckerberg hired Joseph DeRisi, a biochemist at the University of California San Francisco, who had invented the technology that first identified severe acute respiratory syndrome, or SARS, which happened to be a coronavirus. Just months before the pandemic hit, Zuckerberg had livestreamed a discussion with DeRisi that touched on advances in virology and addressed "the erosion of a sense of truth and trust in experts."

Facebook could hardly have been better positioned when all hell broke loose. Zuckerberg instructed his lieutenants to begin preparing for a pandemic as early as January, and mandated the Integrity team to begin working on misinformation about COVID. He also wrote to Anthony Fauci with a proposal to create a "COVID Information Center" stocked with authoritative information about the virus and potential content from the director of the National Institute of Allergy and Infectious Diseases at the National Institutes of Health himself. "Don't feel the need to respond to this if it doesn't seem helpful," Zuckerberg wrote.

Like a lot of what Zuckerberg was doing, the suggestion *was* helpful. Fauci took him up on it, and Zuckerberg declared the company would not tolerate bad information that put people at imminent risk of danger.

As Facebook's platform swelled with fundraisers for furloughed workers, restaurants promoting pickup meals, and live-broadcast musical performances, the company won a grudging public reappraisal. As a self-professed wartime CEO, Zuckerberg was perfectly suited to a virus as an adversary. "The Pandemic Is Giving Zuckerberg a Shot at Making Amends," *Bloomberg Businessweek* announced. A *New York Times* feature said that, after years of mistakes, the coronavirus had given the CEO the "opportunity to demonstrate that he has grown into his responsibilities as a leader."

Internally, Facebook staff seemed to agree. Facebook polled its employees on an ongoing basis and found that, even amid the chaos of the early pandemic, employee sentiment was surging. On May 25, 83 percent of the company's staff reported optimism about Facebook, a rise of more than 25 points in two months. Nearly the same percentage reported confidence in the company's leadership, a level not recorded since before the 2016 election.

There were signs, however, that all was not well. In May, a data scientist working on integrity posted a Workplace note titled "Facebook Creating a Big Echo Chamber for 'the Government and Public Health Officials Are Lying to Us' Narrative—Do We Care?"

Just a few months into the pandemic, groups devoted to opposing COVID lockdown measures had become some of the most widely viewed on the platform, pushing false claims about the pandemic under the guise of political activism. Beyond serving as an echo chamber for alternating claims that the virus was a Chinese plot and that the virus wasn't real, the groups served as a staging area for platform-wide assaults on mainstream medical information. Doctors describing packed emergency rooms and refrigerated trucks were getting shouted down; expressions of doubt colonized news stories about public health guidance like barnacles encrusting a pier.

An analysis showed these groups had appeared abruptly, and

while they had ties to well-established anti-vaccination communities, they weren't arising organically. Many shared near-identical names and descriptions, and an analysis of their growth showed that "a relatively small number of people" were sending automated invitations to "hundreds or thousands of users per day."

Most of this didn't violate Facebook's rules, the data scientist noted in his post. Claiming that COVID was a plot by Bill Gates to enrich himself from vaccines didn't meet Facebook's definition of "imminent harm." But, he said, the company should think about whether it was merely reflecting a widespread skepticism of COVID or creating one.

"This is severely impacting public health attitudes," a senior data scientist responded. "I have some upcoming survey data that suggests some baaaad results."

Soon afterward, the company declared a "site event," or SEV, formally recognizing a significant problem. The emergency in question was that misinformation appeared to be up by more than double.

Misinformation regarding COVID-19 spreading from QAnon and other conspiracy movements wasn't the only problem about to overtake Facebook.

President Trump was gearing up for reelection and he took to his platform of choice, Twitter, to launch what would become a monthslong attempt to undermine the legitimacy of the November 2020 election. "There is no way (ZERO!) that Mail-In Ballots will be anything less than substantially fraudulent," Trump wrote. As was standard for Trump's tweets, the message was cross-posted on Facebook.

Under the tweet, Twitter included a small alert that encouraged users to "Get the facts about mail-in ballots." Anyone clicking on it was informed that Trump's allegations of a "rigged" election were false and there was no evidence that mail-in ballots posed a risk of fraud.

Twitter had drawn its line. Facebook now had to choose where it stood. Monika Bickert, Facebook's head of Content Policy, declared that Trump's post was right on the edge of the sort of misinformation about "methods for voting" that the company had already pledged to take down.

Zuckerberg didn't have a strong position, so he went with his gut and left it up. But then he went on Fox News to attack Twitter for doing the opposite. "I just believe strongly that Facebook shouldn't be the arbiter of truth of everything that people say online," he told host Dana Perino. "Private companies probably shouldn't be, especially these platform companies, shouldn't be in the position of doing that."

The interview caused some tumult inside Facebook. Why would Zuckerberg encourage Trump's testing of the platform's boundaries by declaring its tolerance of the post a matter of principle?

The perception that Zuckerberg was kowtowing to Trump was about to get a lot worse. On the day of his Fox News interview, protests over the recent killing of George Floyd by Minneapolis police officers had gone national, and the following day the president tweeted that "when the looting starts, the shooting starts"—a notoriously menacing phrase used by a white Miami police chief during the civil rights era.

Declaring that Trump had violated its rules against glorifying violence, Twitter took the rare step of limiting the public's ability to see the tweet—users had to click through a warning to view it, and they were prevented from liking or retweeting it.

Over on Facebook, where the message had been cross-posted as usual, the company's classifier for violence and incitement estimated it had just under a 90 percent probability of breaking the platform's rules—just shy of the threshold that would get a regular user's post automatically deleted.

Trump wasn't a regular user, of course. As a public figure, arguably the world's *most* public figure, his account and posts were protected by dozens of different layers of safeguards.

Facebook had developed the concept of applying different standards of treatment to public figures in a way that has become the stuff of company lore. During Barack Obama's first term as president, he had posted a list of his favorite books, including Herman

Melville's celebrated novel about obsession, industrial progress, and a white whale. The title tripped the wire of an early effort to automate the removal of obscenities.

The resulting outage of Obama's page was brief and ultimately inconsequential, but it sent a message to Facebook's leadership: the company could not get away with botching moderation calls on the president of the United States forever. The incident and others like it led to the concept of "shielding," which was later renamed XCheck (pronounced "cross-check").

To implement it, Facebook drew up a list of accounts that were immune to some or all immediate enforcement actions. If those accounts appeared to break Facebook's rules, the issue would go up the chain of Facebook's hierarchy and a decision would be made on whether to take action against the account or not. Every social media platform ended up creating similar lists—it didn't make sense to adjudicate complaints about heads of state, famous athletes, or persecuted human rights advocates in the same way the companies did with run-of-the-mill users. The problem was that, like a lot of things at Facebook, the company's process got particularly messy.

For Facebook, the risks that arose from shielding too few users were seen as far greater than the risks of shielding too many. Erroneously removing a bigshot's content could unleash public hell—in Facebook parlance, a "media escalation" or, that most dreaded of events, a "PR fire." Hours or days of coverage would follow when Facebook erroneously removed posts from breast cancer victims or activists of all stripes. When it took down a photo of a risqué French magazine cover posted to Instagram by the American singer Rihanna in 2014, it nearly caused an international incident. As internal reviews of the system later noted, the incentive was to shield as heavily as possible any account with enough clout to cause undue attention.

No one team oversaw XCheck, and the term didn't even have a specific definition. There were endless varieties and gradations applied to advertisers, posts, pages, and politicians, with hundreds of engineers around the company coding different flavors of protections and tagging accounts as needed. Eventually, at least 6 million accounts and pages were enrolled into XCheck, with an internal guide stating that an entity should be "newsworthy," "influential or popular," or "PR risky" to qualify. On Instagram, XCheck even covered popular animal influencers, including Doug the Pug.

Any Facebook employee who knew the ropes could go into the system and flag accounts for special handling. XCheck was used by more than forty teams inside the company. Sometimes there were records of how they had deployed it and sometimes there were not. Later reviews would find that XCheck's protections had been granted to "abusive accounts" and "persistent violators" of Facebook's rules.

The job of giving a second review to violating content from high-profile users would require a sizable team of full-time employees. Facebook simply never staffed one. Flagged posts were put into a queue that no one ever considered, sweeping already once-validated complaints under the digital rug. "Because there was no governance or rigor, those queues might as well not have existed," recalled someone who worked with the system. "The interest was in protecting the business, and that meant making sure we don't take down a whale's post."

The stakes could be high. XCheck protected high-profile accounts, including in Myanmar, where public figures were using Facebook to incite genocide. It shielded the account of British far-right figure Tommy Robinson, an investigation by Britain's Channel Four revealed in 2018.

One of the most explosive cases was that of Brazilian soccer star Neymar, whose 150 million Instagram followers placed him among the platform's top twenty influencers. After a woman accused Neymar of rape in 2019, he accused the woman of extorting him and posted Facebook and Instagram videos defending himself—and showing viewers his WhatsApp correspondence with his accuser, which included her name and nude photos of her. Facebook's procedure for handling the posting of "non-consensual intimate imagery" was simple: delete it. But Neymar was protected by XCheck. For more than a day, the system blocked Facebook's moderators from removing the video. An internal review of the incident found that 56 million Facebook and Instagram users saw what Facebook described in a separate document as "revenge porn," exposing the woman to what an employee referred to in the review as "ongoing abuse" from other users.

Facebook's operational guidelines stipulate that not only should unauthorized nude photos be deleted, but people who post them should have their accounts deleted. Faced with the prospect of scrubbing one of the world's most famous athletes from its platform, Facebook blinked.

"After escalating the case to leadership," the review said, "we decided to leave Neymar's accounts active, a departure from our usual 'one strike' profile disable policy."

Facebook knew that providing preferential treatment to famous and powerful users was problematic at best and unacceptable at worst. "Unlike the rest of our community, these people can violate our standards without any consequences," a 2019 review noted, calling the system "not publicly defensible."

Nowhere did XCheck interventions occur more than in American politics, especially on the right. As Trump continued to rage against unfriendly media coverage, conservatives increasingly started to deploy accusations of bias, turning run-of-the-mill complaints about social media's moderation practices into major scandals. In one instance in 2018, Facebook limited the spread of posts by pro-Trump social media personalities Diamond and Silk as a result of repeated infractions. The company had sent the two an email stating that their content was "unsafe to the community," declaring "this decision is final and it is not appeal-able in any way."

After Senator Ted Cruz accused Zuckerberg of censoring the women, Facebook declared that its message to the duo had been "inaccurate" and lifted the penalty. That wasn't enough to calm the waters, though. Within the month, Diamond and Silk were testifying in front of the Republican-controlled House Judiciary Committee.

Such incidents led to another form of coddling, known as the "self-remediation window." When a high-enough-profile account was conclusively found to have broken Facebook's rules, the company would delay taking action for twenty-four hours, during which it tried to convince the offending party to remove the offending post voluntarily. The program served as an invitation for privileged accounts to play at the edge of Facebook's tolerance. If they crossed the line, they could simply take it back, having already gotten most of the traffic they would receive anyway. (Along with Diamond and Silk, every member of Congress ended up being granted the self-remediation window.)

Sometimes Kaplan himself got directly involved. According to documents first obtained by BuzzFeed, the global head of Public Policy was not above either pushing employees to lift penalties against high-profile conservatives for spreading false information or leaning on Facebook's fact-checkers to alter their verdicts.

An understanding began to dawn among the politically powerful: if you mattered enough, Facebook would often cut you slack. Prominent entities rightly treated any significant punishment as a sign that Facebook didn't consider them worthy of white-glove treatment. To prove the company wrong, they would scream as loudly as they could in response.

"Some of these people were real gems," recalled Harbath. In Facebook's Washington, DC, office, staffers would explicitly justify blocking penalties against "Activist Mommy," a Midwestern Christian account with a penchant for anti-gay rhetoric, because she would immediately go to the conservative press.

Facebook's fear of messing up with a major public figure was so great that some achieved a status beyond XCheck and were whitelisted altogether, rendering even their most vile content immune from penalties, downranking, and, in some cases, even internal review.

This was the ocean of special treatment that Trump was swimming in when he tweeted about looting and shooting. The stakes could not be higher. The question of what to do would go to Zuckerberg.

Members of Civic watched the controversy engulf the platform in real time. In the immediate aftermath of Floyd's murder, which was caught on video, a map of hate speech and negative user feedback across the United States showed a sea of green, nothing above baseline. After protests in Minneapolis turned violent, pockets of red began popping up in and around the city, but nowhere else. After Trump's Facebook followers began resharing his message—often into groups where the content became a magnet for users' most violent and bigoted opinions—most of America's major cities turned bright red.

"At the end of June 2, we can see clearly that the entire country was basically 'on fire,'" wrote four data scientists in a later analysis titled "Hate Begets Hate; Violence Begets Violence."

There would be plenty of time in coming weeks to consider the interpretation of Trump's words, their historical context, and the president's intent. Cable TV hosts would rehash these questions ad infinitum, and Facebook employees would debate how the platform's thicket of community standards and implementation guidelines applied. But intentions and rules were profoundly beside the point.

What mattered, as the data scientists had noted, was that the post behaved like hate speech. The prospect of answering property crimes with lethal force functioned as hate bait on a national scale, launching a public conversation that made people want to claw each other's eyes out, creating "a high degree of correlation between the user-reported harm online and the offline protest events."

The data scientists couldn't establish causality, but they could clearly observe that the spread of the material was not random. No sooner had Trump's cross-posted tweet appeared on Facebook than a "set of distributors" began sharing it into partisan networks where it spurred even more inflammatory responses featuring phrases like "real bullets," "fuck the white . . . ," and "shooting the bastards."

This was not surprising to Facebook's technocrats. Earlier At Risk Countries work had established that such incendiary material provided a scaffolding for hate the way a coral reef builds up around a sunken ship.

"Hate and violence can be exacerbated by groups distribution," the authors of the "Hate Begets Hate" note summarized, and the activity around Trump's post provided an American analogue to previous findings that such activity fed "intercommunal violence."

Stopping short of restricting the sharing of Trump's post or removing it, there were steps that the company could have taken to limit the damage. It could have limited reshares, lowered the precision required to hide edgy comments, or asked users to reconsider spicy language used in discussing the post.

Zuckerberg decided instead to do nothing, declaring that Trump's comment amounted to an informational warning about the potential use of state violence and questionably telling employees that the phrase had no history as a racist dog whistle. The reaction that followed was swift and angry, both inside and outside the company.

The job of internally explaining Zuckerberg's decision fell to Bickert, and it wasn't an easy one. "How can we look our employees in the eye and say we're okay with this?" asked Mike Schroepfer, Facebook's Chief Technology Officer. (In later public interviews, Schroepfer was more circumspect, saying only that it was important that the company make clear rules and follow them.)

The conclusions of the "Hate Begets Hate" note were bleak enough that the authors appended a postscript addressed "To Our Colleagues." That Facebook now had the knowledge and tools to do better but hadn't was "disheartening." But "when lives are on the line, not trying is not an option," the postscript read, quoting Ime Archibong, one of Facebook's most senior Black executives. "Stay, and fight the good fight together."

In the aftermath of looting and shooting, the already freewheeling tone of Workplace conversation kicked into a higher gear. Some of the posts were combative denunciations of Zuckerberg's leadership; others were geared toward not letting a crisis go to waste.

Chakrabarti wrote a post titled "Bending Our Platforms Toward Racial Justice," approaching the subject in a measured way but making proposals that were nothing short of revolutionary. The company's platform, he wrote, wasn't built for functional public debate and was currently "prioritizing regulatory interests over community protection" while failing to "support pro-democracy values."

In a jab at the Public Policy team with which he had been wrangling for years, he wrote that fixing the problem would require separating Facebook's Content Policy decisions from its Washington lobbying operation as well as giving the public and employees a say in platform governance decisions that Zuckerberg was making. In a concurring post, Iyer, the cohead of Civic alongside Chakrabarti, wrote that inaction meant Facebook "will fail at bringing the company along with the decisions Mark makes this year."

A chorus of voices soon chimed in in agreement. "Great post as usual," wrote Fidji Simo, the head of the Facebook app. "What can I do to help advocate?" asked Margaret Stewart, vice president of Product Design. "Whatever tools we have are at your service," wrote Miranda Sissons, Facebook's director of Human Rights. "Count me in to support and help in any way I can," wrote Kristin Hendrix, Instagram's head of Research.

The post received a decidedly cooler response from Chakrabarti's boss, Guy Rosen, who privately scolded him that "such talk did not befit a leader at the company."

Boland, the vice president of Partnerships who had been with the company since 2009, was less concerned by the rhetoric from the president than with what he was seeing on CrowdTangle. There was no way to be proud of the content that the platform was amplifying. He called Sandberg and told her that, after eleven years at Facebook, he was going to quit.

"You shouldn't do that right now," Sandberg said, then let Boland in on a secret: Chris Cox was coming back to Facebook. The problems Boland saw were real, she said, and rather than quitting, he should bring them up with Cox in a few weeks. Boland agreed to give it a shot.

Outside the executive ranks, the mood among employees was as close to mutinous as it had ever been. Four days after Trump's tweet, employees staged a virtual walkout. In a rare turn of events, many started voicing their concerns outside the confines of the company's internal chatrooms, taking their dissatisfaction to Twitter. "Mark is wrong, and I will endeavor in the loudest possible way to change his mind," tweeted Ryan Freitas, a director who oversaw the product design of News Feed and its associated integrity work.

At a company Q&A, a member of the Integrity team asked Zuckerberg why Rosen, whose team was responsible for societal violence concerns, hadn't been involved in the decision to leave Trump's post untouched.

"I'm sorry. I believe Guy was included," Zuckerberg replied. That wasn't accurate, and the Integrity staffer knew it. Rosen had told his entire team that morning that he was not consulted. The staffer was ready to go in for the kill.

"So I attended Guy's Q&A this morning—" she began.

Zuckerberg then appeared to waver. "Maybe he wasn't," Zuckerberg said of Rosen's involvement. "I actually—I'm not sure if he was."

That Zuckerberg couldn't recall whether he had consulted Facebook's VP of Integrity on an integrity decision wasn't "great," the staffer declared. With the CEO on the defensive, she asked him to commit to involving the team that handled societal violence risks when making such decisions in the future.

That's when her connection to the Zoom call cut out.

"Sorry, I lost you there," Zuckerberg said, assuring his remaining audience that "all arguments were considered."

Nobody could confidently say the questioner's disconnection hadn't been an accident. But her colleagues' suspicion that Zuckerberg's interrogation had been unceremoniously cut short was later buttressed by a change in policy.

Facebook town halls had a long tradition of frank exchanges and few leaks. In 2017, Bloomberg had published a story marveling at Facebook leaders' ability to have forthright, company-wide discussions without juicy details leaking out. That tradition was now over. Hours after the Q&A concluded, *Vox* published the full transcript. In the wake of the meeting, Facebook announced that future questions at company town halls would be prerecorded. (The company later attributed the change to the pandemic and noted that employees can still try to ask live questions in the comments section.)

Leaks nonetheless became a near-weekly occurrence, feeding a deluge of articles about the company's failures and how it exercised favoritism in enforcing the rules it did have. The company had repeatedly overridden fact-checks for prominent conservative publishers, including Breitbart, *BuzzFeed News* reported, with Kaplan personally intervening to shield the account of right-wing firebrand Charlie Kirk.

At the end of August, less than one month after that story was published, a baby-faced white kid named Kyle Rittenhouse drove from his home in Indiana to Kenosha, Wisconsin, where civil unrest had broken out following the police shooting of a Black man named Jacob Blake. Once there, Rittenhouse shot two people to death and maimed a third. He had taken the trip after a local man created a Facebook event calling for volunteers to "take up arms and defend out [*sic*] City tonight from the evil thugs." The post, which was also amplified by radio and other media as it began growing in popularity, had been flagged by Facebook users 455 times. Zuckerberg pronounced the company's failure to remove the event "an operational mistake."

Members of Integrity were left scratching their heads after Zuckerberg's declaration that Facebook had simply erred in its application of platform rules. Reality was much messier. "The language used by these events and groups was not violating," a research manager for Civic Integrity rightly noted on Workplace, further predicting that Facebook's attempt to separate out praise for Rittenhouse's actions (which was banned) from questioning whether he was being treated fairly (which was allowed) would be a disaster. The company didn't remotely have the enforcement capabilities to do that right, and it probably never would.

Other Civic colleagues and Integrity staffers piled into the comments section to concur. "If our goal, was say something like: have less hate, violence etc. on our platform to begin with instead of remove more hate, violence etc. our solutions and investments would probably look quite different," one wrote.

Rosen was getting tired of dealing with Civic. Zuckerberg, who famously did not like to revisit decisions once they were made, had already dictated his preferred approach: automatically remove content if Facebook's classifiers were highly confident that it broke the platform's rules and take "soft" actions such as demotions when the systems predicted a violation was more likely than not. These were the marching orders and the only productive path forward was to diligently execute them.

Other people who worked on integrity trusted the decisions of colleagues elsewhere in the company, but Civic staffers seemed to think they knew better, Rosen told him. Civic staffers were citing the team's oath in meetings, and Rosen found the implication that Civic was adhering to a higher ethical standard than the rest of the company to be toxic. Chakrabarti should never have created the principles in the first place, he said, and warned Chakrabarti that Civic would have to be "zero drama" for him to have a shot at even a middling performance review.

12

It was August 2020, and Kiran, a staffer at Facebook India, was about to do something that was personally unwise.

The week before, the *Wall Street Journal* had published a story my colleague Newley Purnell and I cowrote about how Facebook had exempted a firebrand Hindu politician from its hate speech enforcement. There had been no question that Raja Singh, a member of the Telangana state parliament, was inciting violence. He gave speeches calling for Rohingya immigrants who fled genocide in Myanmar to be shot, branded all Indian Muslims traitors, and threatened to raze mosques. He did these things while building an audience of more than 400,000 followers on Facebook. Earlier that year, police in Hyderabad had placed him under house arrest to prevent him from leading supporters to the scene of recent religious violence.

That Facebook did nothing in the face of such rhetoric could have been due to negligence—there were a lot of firebrand politicians offering a lot of incitement in a lot of different languages around the world. But in this case, Facebook was well aware of Singh's behavior. Indian civil rights groups had brought him to the attention of staff in both Delhi and Menlo Park as part of their efforts to pressure the company to act against hate speech in the country.

There was no question whether Singh qualified as a "dangerous individual," someone who would normally be barred from having a presence on Facebook's platforms. Despite the internal conclusion that Singh and several other Hindu nationalist figures were creating a risk of actual bloodshed, their designation as hate figures had been blocked by Ankhi Das, Facebook's head of Indian Public Policy—the same executive who had lobbied years earlier to reinstate BJP-associated pages after Civic had fought to take them down.

Das, whose job included lobbying India's government on Facebook's behalf, didn't bother trying to justify protecting Singh and other Hindu nationalists on technical or procedural grounds. She flatly said that designating them as hate figures would anger the government, and the ruling BJP, so the company would not be doing it.

My colleague Newley Purnell and I had learned of Das's protection of Singh, and we'd written it up as both an apparent violation of Facebook's pledge to safeguard users and an extreme example of the company's suspected willingness to bend the rules on behalf of the powerful. Such an open declaration of subservience to a government would have been unthinkable from a senior Facebook executive in the U.S., but India was obviously different.

Following our story, Facebook India's then–managing director Ajit Mohan assured the company's Muslim employees that we had gotten it wrong. Facebook removed hate speech "as soon as it became aware of it" and would never compromise its community standards for political purposes. "While we know there is more to do, we are making progress every day," he wrote.

It was after we published the story that Kiran (a pseudonym) reached out to me. They wanted to make clear that our story in the *Journal* had just scratched the surface. Das's ties with the government were far tighter than we understood, they said, and Facebook India was protecting entities much more dangerous than Singh.

Kiran was prepared to elaborate on these things, but not provide their name, not even on encrypted channels. The risk wasn't just that they would get fired, they said. India was no longer a place where speaking out about the government and its use of Facebook was a safe thing to do.

Kiran shared a few details of their history. They had become smitten with Facebook while at university, with the platform sweeping college campuses across India. That trendiness wasn't Facebook's central appeal for them, however. Kiran belonged to a marginalized group, and the platform had allowed them to find a community that was far less accessible offline.

Their hiring by Facebook—to work on safety issues with teams in both India and the United States—felt like a triumph. "I got in because I genuinely loved Facebook," they said.

But from the get-go, Kiran could see that something was clearly wrong. Part of the problem was Das's political favoritism. She made sure that Facebook donated to government-aligned charities, and conducted live interviews on Facebook's official pages with BJP ministers looking to raise their profile. She was not shy about internally expressing her support for Modi's government.

Das's colleagues sometimes found it distasteful that she spoke like a BJP campaign operative, but partisan ties weren't unusual for Facebook's head of Public Policy in important markets. Joel Kaplan had Republican affiliations in the United States; he had inflamed colleagues when he showed up to support Brett Kavanaugh during the future Supreme Court justice's congressional hearing on sexual assault allegations against him. In Israel, the head of Policy was Jordana Cutler, a former aide to Prime Minister Benjamin Netanyahu. Facebook wanted friendly relationships with governments, so it hired people who already had them.

But Kiran, whose job exposed them to Facebook's handling of high-profile content moderation issues around the world, saw something darker at play. When American employees warned about public figures using the platform to incite violence, they were talking about Trump cross-posting a menacing, coded one-liner from Twitter. In India, the better reference point was not the U.S. president but what had happened a few months prior, in February 2020, when Hindu activists in Delhi, some with BJP affiliations, had called on their Facebook followers to break a monthslong sit-in strike by Muslims—by force.

"Hindus, come out. Die or kill," one prominent activist had declared during a Facebook livestream, according to a later report by retired Indian civil servants. The ensuing violence left fifty-three people dead and swaths of northeastern Delhi burned.

On a visceral level, Kiran believed, Facebook employees in

the United States didn't grasp how dire the situation in India was becoming. They didn't see the Hindu groups devoted to the protection of cows that crowdsourced rumors of cattle being transported for slaughter and then documented the vigilante justice meted out to truck drivers. They didn't know that the police weren't going to show up when militant Hindu groups raided churches accused of performing conversions. And they didn't realize that Facebook posts about "Love Jihad"—an Islamophobic conspiracy theory that Muslim men were looking to take over India by seducing naive Hindu girls—were getting people killed.

So Kiran tried to explain it to their colleagues in Menlo Park. They participated in policy forums and joined other employees in successfully prodding the company to ban caste-based hatred. They noted how closely the rhetoric of Love Jihad mirrored the portrayals of Black American men as hypersexual predators, slander entwined with many Jim Crow–era lynchings. But while their American colleagues agreed that the company had serious integrity problems in India, little seemed to happen. "Everyone would say yes yes yes, and then it would perpetually be a work in progress," Kiran said.

Kiran's sense that their American colleagues generally failed to grasp the state of Facebook in India turned out to be correct. Employees at Facebook—if they even used the platform regularly outside of work—might as well have been on a different platform.

From the company's point of view, this was an intentional design choice. Ever since Facebook's 2008 decision to hasten its international push by delegating the translation of its site to volunteer users abroad, the platform had largely considered incomprehension of how it was being used a perk of its business model. The assumption that users would simply sort things out themselves perhaps also explained why the company never bothered to translate its community standards into the dozens of languages in which it offered the site, and why it never hired even a single moderator capable of speaking the main language of every country in which it operated.

The cost of this expeditiousness had been ignorance. Even as the company built out data-light versions of its products to run on the

cheap devices favored in poorer markets and paid frontier telecommunications to provide free Facebook access in villages that didn't
have roads, it continued building and optimizing the platform
for people who used it like they did—people who accepted friend
requests only from people they knew, curated what pages they followed, and muted acquaintances who shared too much junk.

For people with engineering degrees, such routine online
hygiene didn't feel like a demonstration of sophistication. They
were, however, skills. The choices users made not only constrained
what Facebook showed them in News Feed. It constrained what the
platform showed their connections, too. Having well-heeled friends
beat even the best misinformation classifier when it came to keeping
conspiracy theories out of one's feed.

Like all forms of herd immunity, the protections weren't perfect.
Everyone has an uncle or childhood friend who might reshare crazy
stuff and bite on engagement bait. But the protective effect of having
educated, internet-savvy friends helped explain why the company's
senior leadership had been surprised to discover that Macedonian
troll farms were producing the platform's top election content. Fires
that broke out on Facebook's lower floors had to become full-on
infernos before its penthouse dwellers ever smelled smoke.

Facebook was paying its Integrity staff to watch for trouble, but
even the professional monitors got caught off guard in markets like
India. Ahead of the Indian elections in 2019, the company had sent
user experience researchers to the country to take the pulse of Indians on Facebook, including the researcher who had chronicled the
nasty tendencies of Facebook's recommendation systems in "Carol's
Journey to QAnon."

The researcher set up a dummy account while traveling. Because
the platform factored a user's geography into content recommendations, she and a colleague noted in a writeup of her findings, it was
the only way to get a true read on what the platform was serving up
to a new Indian user.

Ominously, her summary of what Facebook had recommended
to their notional twenty-one-year-old Indian woman began with

a trigger warning for graphic violence. While Facebook's push of American test users toward conspiracy theories had been concerning, the Indian version was dystopian.

"In the 3 weeks since the account has been opened, by following just this recommended content, the test user's News Feed has become a near constant barrage of polarizing nationalist content, misinformation, and violence and gore," the note stated. The dummy account's feed had turned especially dark after border skirmishes between Pakistan and India in early 2019. Amid a period of extreme military tensions, Facebook funneled the user toward groups filled with content promoting full-scale war and mocking images of corpses with laughing emojis.

This wasn't a case of bad posts slipping past Facebook's defenses, or one Indian user going down a nationalistic rabbit hole. What Facebook was recommending to the young woman had been bad from the start. The platform had pushed her to join groups clogged with images of corpses, watch purported footage of fictional air strikes, and congratulate nonexistent fighter pilots on their bravery.

"I've seen more images of dead people in the past three weeks than I've seen in my entire life, total," the researcher wrote, noting that the platform had allowed falsehoods, dehumanizing rhetoric, and violence to "totally take over during a major crisis event." Facebook needed to consider not only how its recommendation systems were affecting "users who are different from us," she concluded, but rethink how it built its products for "non-US contexts."

India was not an outlier. Outside of English-speaking countries and Western Europe, users routinely saw more cruelty, engagement bait, and falsehoods. Perhaps differing cultural senses of propriety explained some of the gap, but a lot clearly stemmed from differences in investment and concern.

From 2016 onward, "PR fires" in the United States had driven much of Facebook's integrity work. The company built its defenses against foreign interference in response to U.S. election concerns, set privacy policies in the aftermath of Cambridge Analytica, and wrote its rules on vaccine misinformation in response to a California measles outbreak. In every instance, large-scale bad press had driven Facebook's efforts to fix problems. Other large and affluent markets sometimes garnered the same attention.

PR fires tended to smolder less as the countries they occurred in got smaller and poorer—even if the problems had equal or greater human consequences. The places where Facebook invested the least attention and resources were often those where its product was most central to public life.

In 2019, the BBC had run an exposé on the sale of domestic servants in Persian Gulf States via Instagram and bartering apps available through the Apple and Google stores. It was grim stuff: women from Nepal, the Philippines, and countries across Africa lured to Kuwait and Saudi Arabia with false promises of jobs had their passports seized. Once in country, the women were sold from household to household, forced to work seven days a week and subjected to confinement.

This wasn't supposed to be legal in the Gulf under the gray-market labor sponsorship system known as *kafala,* but the internet had removed the friction from buying people. Undercover reporters from BBC Arabic posed as a Kuwaiti couple and negotiated to buy a sixteen-year-old girl whose seller boasted about never allowing her to leave the house.

Everyone told the BBC they were horrified. Kuwaiti police rescued the girl and sent her home. Apple and Google pledged to root out the abuse, and the bartering apps cited in the story deleted their "domestic help" sections. Facebook pledged to take action and deleted a popular hashtag used to advertise maids for sale.

After that, the company largely dropped the matter. But Apple turned out to have a longer attention span. In October, after sending Facebook numerous examples of ongoing maid sales via Instagram, it threatened to remove Facebook's products from its App Store.

Unlike human trafficking, this, to Facebook, was a real crisis.

"Removing our applications from Apple's platforms would have had potentially severe consequences to the business, including depriving millions of users of access to IG & FB," an internal report on the incident stated.

With alarm bells ringing at the highest levels, the company found and deleted an astonishing 133,000 posts, groups, and accounts related to the practice within days. It also performed a quick revamp of its policies, reversing a previous rule allowing the sale of maids through "brick and mortar" businesses. (To avoid upsetting the sensibilities of Gulf State "partners," the company had previously permitted the advertising and sale of servants by businesses with a physical address.) Facebook also committed to "holistic enforcement against any and all content promoting domestic servitude," according to the memo.

Apple lifted its threat, but again Facebook wouldn't live up to its pledges. Two years later, in late 2021, an Integrity staffer would write up an investigation titled "Domestic Servitude: This Shouldn't Happen on FB and How We Can Fix It." Focused on the Philippines, the memo described how fly-by-night employment agencies were recruiting women with "unrealistic promises" and then selling them into debt bondage overseas. If Instagram was where domestic servants were sold, Facebook was where they were recruited.

Accessing the direct-messaging inboxes of the placing agencies, the staffer found Filipina domestic servants pleading for help. Some reported rape or sent pictures of bruises from being hit. Others hadn't been paid in months. Still others reported being locked up and starved. The labor agencies didn't help.

The passionately worded memo, and others like it, listed numerous things the company could do to prevent the abuse. There were improvements to classifiers, policy changes, and public service announcements to run. Using machine learning, Facebook could identify Filipinas who were looking for overseas work and then inform them of how to spot red flags in job postings. In Persian Gulf countries, Instagram could run PSAs about workers' rights.

These things largely didn't happen for a host of reasons. One memo noted a concern that, if worded too strongly, Arabic-language PSAs admonishing against the abuse of domestic servants might "alienate buyers" of them. But the main obstacle, according to people familiar with the team, was simply resources. The team devoted full-time to human trafficking—which included not just the smuggling of people for labor and sex but also the sale of human organs—amounted to a half-dozen people worldwide. The team simply wasn't large enough to knock this stuff out.

At the time of this book's writing, it remains readily discoverable.

The company paid staff to look at and sometimes work on addressing these problems and others, but without the regular scrutiny of story-hunting reporters and brand-conscious advertisers, Facebook often seemed institutionally unable—or unwilling—to muster a consistent capacity to focus on them.

Many of the issues were foundational. Facebook never bothered to translate its community standards into languages spoken by tens of millions. It often used contractors to review content in other languages, and would route posts in Syrian and Iraqi Arabic to contractors in Morocco, for whom they were incomprehensible. An internal review referred to that as a failure to meet the "rock-bottom minimum" standard that contract moderators understand the language of the content they reviewed.

More money certainly would have helped with international integrity work, but staff in Menlo Park came to realize that they faced another problem. At a company that required scale in everything it did, classifiers didn't reliably work across languages. Only image-based classifiers, like ones that scanned for nudity, really scaled globally with any ease. Everything else had to be done on a one-off basis. Maintaining the company's civic classifier, for example, required Facebook to stay on top of not just public officeholders but places, people, and terms of societal importance. Near-constant retraining was essential; once-obscure things—say, the name George Floyd—could become flashpoints overnight.

Civic wasn't the only classifier that needed care and feeding. For integrity alone, there was hate speech, terrorism, misinformation, violence and incitement, borderline hostile speech, and others. If the company wanted those to work in Portuguese, they would have to have labeling and engineering teams building them out separately—and then account for the regional discrepancies between Portugal, Brazil, and Angola.

Building classifier variations that reliably worked all around the world might seem a Herculean task until you remembered that classifiers' precision and recall was often marginal even in American English. Then the problem began to look impossible.

AI was the cornerstone of Facebook's strategy to govern its platform, but that strategy couldn't possibly work outside of a handful of major languages in major markets. India alone had more than twenty official languages. As late as 2021, employees spoke of election-related classifiers in the top two, Hindi and Bengali, as "underdeveloped," "out of date," and "not really catching." Even rudimentary classifiers didn't exist for most of the rest of India's languages, which were natively spoken by more people than English worldwide. And Facebook had no plans to create them.

The absence of tailored AI didn't just deprive people who spoke those languages of whatever improvements an automated content moderation system might have provided. It meant that the version of the platform they experienced would only get worse.

By 2020, the tradeoff between growth and integrity work was well accepted inside Facebook. The company could accurately deny that News Feed promoted hate and lies to boost growth, but it could not say that promoting growth in News Feed didn't boost hate and lies *as a side effect.* This reality led to continued negotiations between growth- and integrity-focused teams. If the company altered News Feed in a way that caused sensationalism to spike globally, it could downrank sensationalism in the United States or Germany to offset it. But for many markets lumped into the category known as "Rest of World," no such response was possible.

The circumstances left the At Risk Country team in a bit of a straitjacket. The more the platform invested in language-specific AI to mitigate societal concerns, the bigger the gap would grow between the markets where such classifiers existed and those where they didn't. "Most of our integrity systems are less effective outside of the United States," wrote Tony Leach, the head of the ARC team in a late-2020 summary of its progress. "Classifiers either don't exist or perform less well," he wrote, and "we lack human support."

Leach's team aspired to one day build U.S.-quality classifiers and interventions for nations at risk of societal violence, but the company wasn't getting there anytime soon. In countries including Myanmar and Yemen, Facebook still lacked even rudimentary classifiers to spot incitements to violence and praise of hate figures, much less the ability to perform targeted interventions in a crisis.

Amid Ethiopia's ethno-regional civil war, Facebook didn't have a hate speech classifier for any of the country's major languages. It ran 3 percent as many fact-checks per user there as it did in the United States and didn't even allow people to report violating content in Oromo, the language of Ethiopia's most populous region.

"We're largely blind to problems on our site," Leach's presentation wrote of Ethiopia.

Facebook employees produced a lot of internal work like this: declarations that the company had gotten in over its head, unable to provide even basic remediation to potentially horrific problems. Events on the platform could foreseeably lead to loss of life and almost certainly did, according to human rights groups monitoring Ethiopia. Meareg Amare, a university lecturer in Addis Ababa, was murdered outside his home one month after a post went viral, receiving 35,000 likes, listing his home address and calling for him to be attacked. Facebook failed to remove it. His family is now suing the company.

As it so often did, the company was choosing growth over quality. Efforts to expand service to poorer and more isolated places would not wait for user protections to catch up, and, even in countries at "dire" risk of mass atrocities, the At Risk Countries team needed approval to do things that harmed engagement.

The company's pledge to "fight for free expression around the world," as Zuckerberg had put it in his Georgetown speech, also had its limits. In early 2020, seeking to force Facebook to comply with an order to censor "anti-state" posts, the Vietnamese government had blockaded the company's local servers in the country. The move didn't knock Facebook's services offline; it just rendered the service slow and intermittent.

Facebook held out for seven weeks. Then it acceded to the government's demands, removing the accounts of activists, who were eventually sentenced to lengthy jail terms.

Facebook told Reuters, which broke the story of the company's capitulation, that it still considered free expression to be a fundamental right. But so was using Facebook. "We have taken this action to ensure our services remain available and usable for millions of people in Vietnam, who rely on them every day," the company's statement read.

Despite the constraints, even jaundiced employees could see the value of their work when hate networks were shut down, policies revised, and ranking changes made. The company's scale amplified their accomplishments. An afternoon spent updating Facebook's internal list of Arabic-language anti-LGBTQ slurs had a greater impact on societal discourse than a local nonprofit could dream of.

But doing that work required accepting business-related limits. The dissonance between aspirations and reality was perhaps greatest for employees working on India, which simultaneously posed "tier one" risks of societal violence and also had a government that Facebook feared.

Documents and transcripts of internal meetings among the company's American staff show employees struggling to explain why Facebook wasn't following its normal playbook when dealing with hate speech, the coordination of violence, and government manipulation in India. Employees in Menlo Park discussed the BJP's promotion of the "Love Jihad" lie. They met with human rights organizations that documented the violence committed by the platform's cow-protection vigilantes. And they tracked efforts by the Indian government and its allies to manipulate the platform via networks of accounts. Yet nothing changed.

"We have a lot of business in India, yeah. And we have connections with the government, I guess, so there are some sensitivities around doing a mitigation in India," one employee told another about the company's protracted failure to address abusive behavior by an Indian intelligence service.

During another meeting, a team working on what it called the problem of "politicized hate" informed colleagues that the BJP and its allies were coordinating both the "Love Jihad" slander and another hashtag, #CoronaJihad, premised on the idea that Muslims were infecting Hindus with COVID via halal food.

The Rashtriya Swayamsevak Sangh, or RSS—the umbrella Hindu nationalist movement of which the BJP is the political arm—was promoting these slanders through 6,000 or 7,000 different entities on the platform, with the goal of portraying Indian Muslims as subhuman, the presenter explained. Some of the posts said that the Quran encouraged Muslim men to rape their female family members.

"What they're doing really permeates Indian society," the presenter noted, calling it part of a "larger war."

A colleague at the meeting asked the obvious question. Given the company's conclusive knowledge of the coordinated hate campaign, why hadn't the posts or accounts been taken down?

"Ummm, the answer that I've received for the past year and a half is that it's too politically sensitive to take down RSS content as hate," the presenter said.

Nothing needed to be said in response.

"I see your face," the presenter said. "And I totally agree."

If standing down on coordinated genocidal rhetoric was a hard thing for American employees at Menlo Park to stomach, it was a hell of a lot harder for Kiran.

As a Facebook employee, their future was bright in India. As a person committed to secular democracy, however, they were getting scared. And the latter ended up outweighing the former.

One incident in particular, involving a local political candidate, stuck out. As Kiran recalled it, the guy was a little fish, a Hindu nationalist activist who hadn't achieved Raja Singh's six-digit follower count but was still a provocateur. The man's truly abhorrent behavior had been repeatedly flagged by lower-level moderators, but somehow the company always seemed to give it a pass.

This time was different. The activist had streamed a video in which he and some accomplices kidnapped a man who, they informed the camera, had killed a cow. They took their captive to a construction site and assaulted him while Facebook users heartily cheered in the comments section.

The video was removed and the activist's account flagged for deletion, Kiran said. But the bleakness of watching a person tortured for social media engagement was hard to shake. Beyond the vileness of its contents, the video reflected Kiran's fear that Facebook was changing Indian users' real-world offline behavior.

Facebook's methods of connecting people had, in well under a decade, helped normalize speech that made India's once-strict laws against "promoting enmity" between different races, religions, and castes seem quaint. Seeing Facebook cow vigilante rhetoric acted out still had the ability to shock, but that was perhaps temporary.

A few weeks later, Kiran discovered that the cow vigilante's account had been reinstated. They never learned why.

"This was a time when our team was severely short-staffed," they said. "You couldn't spend a day following up on a case because you had fifty other cases to prioritize."

As a reporter, I'm happy to work with reliable sources regardless of their motivations. Information from someone snubbed for a promotion can be every bit as valuable as that from a whistleblower guided by conscience, and there are probably more of the former than the latter.

The idealistic ones—like Kiran—tend to be harder to protect. A person who believes they have been mistreated by a boss generally doesn't do things that harm their personal interests further. Someone motivated by the conclusion that their employer is accelerating their nation toward majoritarian violence is a lot more prone to taking risks.

At the time I spoke to Kiran, I didn't understand Facebook's internal tracking of employee activity on Workplace well enough to evaluate their personal exposure. They were an incredible source, explaining both the politics of Facebook's India operation and the dynamics of its products overseas. Information from them provided the bulk of two front-page stories. One was about Facebook's internal work related to the Bajrang Dal, an RSS-affiliated youth group. Members of the organization—which includes thousands of local clubs across India—have earned a reputation for violence, ranging from an annual assault of secular couples celebrating Valentine's Day to the murders of Muslims and Christians.

The U.S. government considered the Bajrang Dal a militant organization, and so did Facebook. Bajrang Dal members' habit of livestreaming acts of violence alone made designating the group as a "dangerous organization" an easy call.

But Facebook hadn't acted. One reason was that, like the RSS, the organization had a close relationship with the BJP. Another was that the Bajrang Dal and its allies were sufficiently violent that Facebook's Security team had put in writing that it feared they might physically attack the company's Indian offices. The story illustrated the entrenchment of majoritarian violence on Facebook and the untouchability of the forces behind it. The Bajrang Dal couldn't be labeled a "dangerous organization" because the organization was too dangerous.

Kiran was also the key source for a follow-up story on Ankhi Das. Facebook had always professed its global commitment to keeping the platform absolutely neutral on politics. The company's role was to provide a forum for competition—as Nick Clegg, the former British deputy prime minister now serving as vice president of Global Affairs, put it, not to "pick up a racket and start playing."

Das was not known by colleagues for being a spectator. Still, anonymous allegations of favoritism within Facebook were not enough to justify a news story. Kiran, however, knew where they could find outright proof of bias: Workplace.

The platform, with its Facebook-like features, lulled people into speaking casually. Over the years, Das had boasted about her access to the BJP's senior leadership, insulted rival parties, and chastised her staff for not expressing adequate enthusiasm about Modi's government.

"It's taken thirty years of grassroots work to rid India of state socialism," she wrote after his first election as prime minister, praising his "strongman" tendencies. She even claimed to colleagues that Facebook's assistance with Modi's social media campaign was responsible for his election.

The posts were damning, but they were also a few years old. This was material that no one would be accessing in 2020 unless they were digging. That kind of excavation came at a high risk. It was the sort of thing that Facebook could go back and look for, if such an allegation was published in the *Journal*. Depending on who else had accessed the posts, Kiran might be safe—or they might be outing themselves to corporate security. There was no way to quantify the risk.

Kiran gave me the go-ahead to run the posts anyway. On August 21, 2020, the *Journal* published a story about how the company's Indian Public Policy head was in the tank for the BJP.

Das, who had already apologized to colleagues for her remarks about Muslims, spent some time out of the office. She formally left Facebook six weeks later "to pursue her interest in public service," according to the company.

For reasons of their own security, Kiran's story has to end there, too.

13

The window for Facebook to take meaningful action ahead of the 2020 presidential election was narrowing by the day.

Civic knew that if there was one thing to prioritize, it would be getting groups under control. Groups could be wonderful, of course, particularly if associated with real-world communities or narrow interests (think woodworkers or spouses of people with Alzheimer's). But they could also be very, very bad, particularly where issues like health and politics were concerned. There were "natural cure" groups, with membership numbers in the six figures, convincing people with breast cancer to skip medical treatment and apply a flesh-killing ointment called "black salve" instead. Massive political groups were dominated by small bands of users pushing recycled memes, borderline hate speech, and commercial spam.

Groups had gotten so terrible through a combination of extreme neglect and steroidal growth. One of the product's greatest vulnerabilities was that it allowed committed users to almost single-handedly provoke viral growth. A lone user could, and did, issue 400,000 invitations to QAnon groups over the course of six months. Combined with Facebook's lax criteria for "joining" a group—all a user had to do was interact with one of its posts in News Feed in any fashion—this was a nightmare in the making, especially ahead of a pandemic-era election already drowning in conspiracy theories.

As far back as 2017, Integrity staff had documented how moderately problematic "gateway" groups served as recruitment grounds for malevolent, invite-only groups that organized campaigns to harass users they didn't like. Such groups screened users for possible snitches, sometimes requiring users to use hate speech as a criterion for entry. (This was moderately effective: on more than one occasion, I was kept out of a group by a demand that I use the N-word.)

Two years later, the Integrity team devoted to groups tried to rein in the product, spurred by the existence of a Sri Lankan hate group that had grown to well over 300,000 within three days and the success of anti-vaccination activists angered by an effort to stop a measles outbreak at the time. Naming their initiative "Hunting Invite Whales," staffers with the Groups Integrity and At Risk Countries teams proposed that the company drastically slash the existing limit on group invites, which was then one thousand per account per hour.

A familiar dance ensued. In short: the Groups Product team had growth targets to hit and the proposal was denied.

In an unfortunate coincidence, the focus on the ways that groups' viral growth tools could be misused came just as Zuckerberg was publicly refocusing Facebook's platform to push users toward more private forms of communication and *especially* groups.

The product would be "at the heart of the app," Zuckerberg told me in an April 2019 interview granted to announce the shift. Having recently joined a 96,000-member group myself premised on the claim that cancer victims should reject standard medical advice in favor of "natural cures," I asked Zuckerberg about safety. The CEO was reassuring. "If people really seek it out on their own, fine," he said. But Facebook certainly wouldn't be helping such conspiracy communities grow. The company's big push on groups had been delayed by six months, he said, because Facebook "prioritized a lot of these safety issues."

Despite Zuckerberg's reassurances, the problems with groups remained obvious and chronic. Emails later turned over to House investigators as part of the January 6 Committee investigation show that Samidh Chakrabarti, Guy Rosen, and Tom Alison, then vice president of Engineering, received an assessment that the state of groups in the run-up to the next presidential election was disastrous.

"Harmful civic groups grow faster than our integrity systems can handle," an email to the men stated. "We are still very exposed to risk during US 2020 as these groups proliferate and evade detection."

With the election just months away, Integrity formed a "Groups Task Force" to work on getting company leadership to revisit their hands-off approach. Trump was refusing to say whether he would accept the results of the election if he lost to Joe Biden. Concerns over turmoil, and even violence, were growing.

The task force had two main functions. The first focused on "Abusive Groups Growth," suggesting measures to slow the growth of groups at least long enough to give the team time to address violations. Whatever Facebook's normal tolerance for terrible groups' rapid growth, it should be lowered at least for now, staffers argued. In a presentation to leadership, researchers wrote that roughly "70% of the top 100 most active US Civic Groups are considered non-recommendable for issues such as hate, misinfo, bullying and harassment."

"We need to do something to stop these conversations from happening and growing as quickly as they do," they wrote.

This argument had been made before, and ignored, but a broad swath of Facebook's leadership was getting rattled by the approaching election. In an email to a counterpart in Facebook's Policy department, Rosen said Facebook needed "some brakes" on Groups. "Some of the largest civic groups are literally a couple of weeks old—so imagine these growing fast in October or November before we get our arms around them," Rosen warned.

In a tacit acknowledgment that Facebook's existing oversight of Groups wasn't functional, Sagnik Ghosh and other Civic engineers were drafted to build tools that could reliably surface problem Groups content for human review or assign automated strikes to administrators. None of this stuff was groundbreaking, but in the election's final months, Rosen treated policing high-profile groups as a priority. The same couldn't be said for the company's CEO.

In a bit of irony, just as Facebook's senior managers recognized the toxicity of Groups, the platform removed one of which Zuckerberg was a member. The Group was quickly restored, but not before Zuckerberg launched an internal campaign against social

media overenforcement. Ordering the creation of a team dedicated
to preventing wrongful content takedowns, Zuckerberg demanded
regular briefings on its progress from senior employees. He also suggested that, instead of rigidly enforcing platform rules on content in
Groups, Facebook should defer more to the sensibilities of the users
in them. In response, a staffer proposed entirely exempting private
groups from enforcement for "low-tier hate speech."

This was hardly an auspicious environment to fix what even
Facebook senior managers acknowledged was a glaring problem.
When an engineer suggested that Facebook should lower the num-
ber of strikes required for a Group's takedown from five to three,
News Feed chief John Hegeman responded that he thought the idea
was sound but such a lost cause that it wasn't even worth proposing.
Leadership had already shot down similar proposals, he said, and it
"wasn't super likely" they'd be willing to relitigate the matter.

Despite its intransigent CEO, by September the company had
made a few basic Zuckerberg-approved changes. The company
stopped actively boosting non-recommended groups in News Feed
and finally established some limits on "bulk invites." It agreed to
temporarily stop recommending all civic and health groups, and
imposed a three-week waiting period before newly created groups
on any subject were eligible to be recommended to Facebook users.

Finally, the team made a contribution to Civic Integrity's arsenal
of Break the Glass interventions. If all hell broke loose around Election Day, the company could throw a switch and force the administrators of groups with a history of breaking Facebook's rules to
manually approve all member posts. The change would both slow
down the groups overall and force the admins to be responsible for
the content they approved.

A second part of the Groups Task Force effort focused on finally
dealing with some of the worst groups on the site. In an effort to get
approval to remove them, the team's leader began sending Rosen
and other senior executives daily analyses of activity in the groups
along with samples of their worst content.

The stuff was viscerally terrible—people clamoring for lynchings
and civil war. One group was filled with "enthusiastic calls for vio-

lence every day." Another top group claimed it was set up by Trump-supporting patriots but was actually run by "financially motivated Albanians" directing a million views daily to fake news stories and other provocative content.

The comments were often worse than the posts themselves, and even this was by design. The content of the posts would be incendiary but fall just shy of Facebook's boundaries for removal—it would be bad enough, however, to harvest user anger, classic "hate bait." The administrators were professionals, and they understood the platform's weaknesses every bit as well as Civic did. In News Feed, anger would rise like a hot-air balloon, and such comments could take a group to the top.

Public Policy had previously refused to act on hate bait. How was it fair to blame whoever posted a news article for the comments beneath it? But, given the approaching election and the general state of turmoil the country found itself in, everyone was a little more receptive than they had once been. Posting hate bait still wasn't a violation of Facebook's rules, but leadership agreed that if the Groups Task Force could scrounge up enough examples of official misbehavior by a certain group, then they could shut it down. A team of staffers would monitor known problematic groups, wait for them to step over the line, and then make the case for why the violations warranted levying a strike against them.

One could argue, as Facebook's Public Policy team long had, that such scrutiny reflected bias. The counterpoint was that the platform was already set up for failure.

"We have heavily overpromised regarding our ability to moderate content on the platform," one data scientist wrote to Rosen in September. "We are breaking and will continue to break our recent promises."

For once, the better-safe-than-sorry argument carried the day.

Over the course of a few months, four hundred politics-related groups were taken down in a process that was, by the team's own admission, shy on "methodological grandeur." But it had an impact, shutting down an estimated 1 billion monthly content views from toxic groups.

Separately, the company also backtracked on political ads, announcing in September that it would ban them for a week before and after the election. If false claims of ballot fraud arose, the last thing Facebook wanted was to take money for promoting them.

The company also backed off its reluctance to label speech by political figures. Building on work previously done by Harbath, Facebook prepared notifications on political posts that would first direct users toward outside sources of factual information about voting and then declare the election over when ballots had been counted.

There was another significant late-game change to Meaningful Social Interaction in the works, too. Though Civic staffers had first warned and then documented how MSI's emphasis on comments and reshares was encouraging anger in politics, Facebook had done little to address the problem in the two and a half years since. But in September, the company abruptly accepted Civic's longstanding recommendation that it diminish the reward for posting content that whipped viewers into a rage. Henceforth, the algorithm would stop treating anger emojis as grounds to amplify a post. Facebook also made plans to stop treating bullying comments as a positive distribution signal.

This was a big victory for Civic and the Integrity Ranking teams, albeit a little late in the day. Properly overhauling News Feed's recommendation systems to reduce the virality of "anger-only content" would take weeks, time that the company no longer had. In the meantime, Facebook would impose a quick and dirty version by simply fiddling with the system's outputs.

"This launch is temporary until core models get updated," noted the author of a September document outlining the ranking change's effects. But even the slapdash solution would produce meaningful reductions in misinformation and incitements to violence. (Facebook would later argue that the change had little effect, though the company has kept it in place.)

The sudden flurry of late-stage approvals for long-stalled integrity projects justified a sense that, somewhere far above Civic, people in the corporate hierarchy were getting nervous. But it was the measures that Facebook was hoping it wouldn't need to take that most clearly demonstrated the company was bracing for a possible electoral crash landing.

Though the company's efforts to create Break the Glass interventions had started in the U.S., the work was geared to less-developed countries. Many, such as aggressively restricting reshares, abstaining from recommending newly created groups, and penalizing the distribution of users who repeatedly posted false information, had begun as proposals for permanent, platform-wide integrity improvements that had been shot down on the basis that they reduced engagement or were deemed objectionable by the U.S.-focused Public Policy team. For major markets, these proposals had hit the end of the road.

In At Risk Countries, however, these discarded integrity levers were given a potential lease on life. Unable to rely on classifiers, and with limited capacity to review content in languages such as Burmese, Tamil, or Oromo, Facebook was willing to alter its mechanics in powerful ways that at home it was not.

Many of the tools relegated to these markets looked suspiciously like the ones that KX Jin had called on the company to apply worldwide the year before. The measures relied far less on AI and far more on friction, rendering the platform more stable in ways that didn't require moderation. They slowed down the spread of viral content, replacing it with material directly posted by a user's friends. They capped how quickly Facebook groups could grow, pulled back on recommending content from newly created sources, and curtailed commenting on posts that seemed to be drawing a lot of hateful responses. There was some aggressive downranking, too. Classifiers got exponentially more effective when you were willing to heavily demote a post that had a fifty-fifty chance of being trouble, though many of the most powerful steps didn't require it.

The first large-scale use of Break the Glass measures—BTGs for short—had come in Myanmar, where in 2018 the company had belatedly thrown whatever it could at tamping down ongoing state-supported violence against the Rohingya. BTGs were next deployed in Sri Lanka on Easter 2019, when devastating suicide attacks against churches and hotels by ISIS-inspired terrorists threatened to kick off wider violence.

"We tried to make the Sri Lanka bombings a big learning event," an ARC staffer recalled. It was hard at first to tell how much various measures were accomplishing, because running A/B tests wasn't ethical in an emergency. But the team had been able to run clean tests later, determining that the most powerful tools all involved content ranking. "The reports we started to get back from civil society was that it was saving lives."

Even when the safety benefits were high and the business risks low—nobody was worried about user growth in Yemen or the Syrian government's thoughts on content moderation—BTG tools were still used sparingly. In advance of Myanmar's 2020 election, for example, the company rolled out a Break the Glass measure that limited the spread of reshared content and replaced it with more content from users' friends. The intervention produced an impressive 25 percent reduction in viral inflammatory posts and a 49 percent reduction in viral hoax photos. These gains—which a memo noted would lower the risk that Facebook would "trigger further conflict" in Myanmar— did not reduce the number of sessions Burmese-language users logged on the app. The only cost was a roughly 2 percent loss in engagement as measured by "Meaningful Social Interactions."

Given Facebook's history of contributing to ethnic cleansing in Myanmar, a 2 percent reduction in reactions and comments would seem a small price to pay for greater safety. But even that cost was too high for the company to bear on a permanent basis. "We plan to roll back this intervention after the Myanmar election in November," the October ARC team document noted. After the Burmese military seized power in a coup, the company kept some calming measures in place but lifted the reshare restrictions that had reduced the spread of inflammatory content. As a matter of principle, Zuckerberg still did not like altering the platform in any way that diminished "user value."

The company was even more reluctant to deploy Break the Glass protections at home. Formally designating the United States a "tier one" At Risk Country would rank it alongside countries like Iraq, Yemen, and Syria, and the company generally tried to avoid discussing the effectiveness of its emergency measures in detail even in those countries. Doing so would reveal the company had significant power over problems like hate speech, the kind that executives like Andrew Bosworth—of the infamous "Ugly Truth" memo—preferred to blame on humanity at large.

Still, the temperature of American politics was undeniably rising, with regular skirmishes between protesters and police in Portland, caravans of Trump supporters attempting to force a Biden campaign bus off the road in Texas, and the sitting president hinting he would not accept defeat. In September, Nick Clegg had told *USA Today* that the company had prepared contingency plans but would not be discussing them because doing so "will no doubt elicit a greater sense of anxiety than we hope will be warranted." The company was trying to reassure the public that it was at least prepared to handle a crisis. Still, the United States would remain an unofficial member of Facebook's potentially-failed-states club.

A senior Facebook Communications executive would later lament that he had been unable to prevent the "Break the Glass" name, with its connotations of fire extinguishers and ringing alarm bells, from taking hold. The term evoked justified panic, an unmitigated crisis, and threats to human life. All valid, given the context. But from a public relations standpoint, it was an absolute dog of a name—and perhaps a little too on the nose.

By October, Civic was back to its War Room footing, monitoring the platform at all hours. After years of pushing for a tougher approach to endemic issues like misinformation and polarization, some of their more ambitious proposals were finally seeing the light of day. The time was fraught—never before in recent history had the country felt so on edge because of domestic politics—but those on the team believed Civic was doing what it could to meet the moment.

Only Chakrabarti and perhaps a couple of trusted colleagues

knew Civic was doing it for the last time. The longstanding conflicts between Civic and Facebook's Product, Policy, and leadership
teams had boiled over in the wake of the "looting/shooting" furor,
and executives—minus Chakrabarti—had privately begun discussing how to address what was now unquestionably viewed as a rogue
Integrity operation. Civic, with its dedicated engineering staff, hefty
research operation, and self-chosen mission statement, was on the
chopping block.

In an email later obtained by the January 6 Committee, Rosen
told Hegeman that, as soon as the election was over, the company
was likely to blow up its existing Integrity structure.

"We're exploring a few models in a very, very tight group," Rosen
wrote, adding that Hegeman should "rest assured" that the reshuf-
fling would address the "rocky relationship your team has with Civic
Integrity specifically."

By either late summer or early fall, Rosen had informed Chakrabarti that the team's fate was sealed. Once the election was over,
Civic would be disbanded. Chakrabarti didn't argue, and, as he later
told deputies, he wasn't even sure he disagreed. The relationship
between his team and the company was probably past repair.

Normally forthright with his product managers during weekly
meetings, Chakrabarti wouldn't tell his deputies for months that
Civic was to be broken up. There was an election to worry about.

The election was all-consuming. Zuckerberg also threw some of
his personal fortune into the fight. The same CEO who had argued at
Georgetown that the best way to tamp down on misinformation was
to focus on the authenticity of the speaker appeared to finally recognize that America was in uncharted waters, facing the possibility of
an all-out assault on democracy. That September, as Trump was in
the midst of his campaign to denigrate mail-in ballots—an option
many voters embraced because of the pandemic—Zuckerberg and
his wife pledged to donate $300 million, via two nonprofits, to local
and state election administration. The sum was equivalent to how
much the federal government allocated to states for the election in
total.

Providing funds to strengthen election infrastructure was in keeping with the CEO's conception of Facebook as rigorously neutral but supportive of the democratic process. But there was, separately, some effort to adjust the company's political positioning in the race's final days.

That Facebook was censoring conservatives was becoming a growing right-wing talking point. The fact of the matter was that much of the misinformation that the company found flooding its platform did tend to target a right-wing audience, including material from foreign actors meddling in U.S. politics. The company had spent the previous four years more concerned about upsetting Trump than about upsetting Democrats. There were valid practical reasons for this. With the White House and Senate under Republican control, Democrats couldn't order regulatory scrutiny or an antitrust action the way the president could.

And though its relations with Republicans tended to get more attention, by the time the election was nearing, relations with Democrats had likewise deteriorated, to the point that Facebook's top liaison to Congress, a former aide to then–House Minority Leader Nancy Pelosi, was barred from visiting her office. With polls strongly suggesting that Trump was going to lose, the company was keenly aware that it was in the hole with an important public constituency.

The Biden campaign was already angry with Facebook for tolerating Trump's repeated statements that the election might be stolen from him, according to people who worked on it. The company's plan to place labels with facts about voting on both candidates' posts certainly did not help. Biden's team saw the prominent advisories as casting unwarranted suspicion on their factual posts aimed at turning out supporters, and the Democratic National Committee raised its displeasure with Facebook. It wanted the labels reserved only for Trump's musings about election fraud.

In October, Rosen came to Chakrabarti asking if there was a way for the company to justify removing the voting information labels from the Democratic candidate's posts. That Biden's camp was asking for the change wasn't mentioned, of course; it was just an idea.

Chakrabarti had long disagreed with Facebook's position that the factuality of politicians' claims about voting and elections was none of its business, and he had led the unsuccessful push to change it ahead of Brazil's 2018 elections. But tweaking the rules in October 2020 in response to what he correctly surmised was a request from Biden's people seemed unacceptable. After years of Civic being accused of liberal bias, Chakrabarti later said to two of his deputies, there was a certain irony in telling Rosen that selective labeling in the Democrats' favor wasn't justifiable under Facebook's policies.

Like the rest of the country, the Civic team went into Election Day nervous. Nobody felt ready. The company had adopted some good measures, but no one felt like the platform was where it needed to be. The risk of chaos seemed high.

Unlike the 2018 War Room, this one was, thanks to the pandemic, just a bunch of people sitting at home in front of their computers. Still, everyone was on high alert, and, to their relief, Tuesday began and ended without major incident. There was no violence at the polls, and no large-scale voter suppression efforts. No last-minute hacked documents, deep fakes, or wildly viral fake news took over the internet.

The Civic team could breathe easy—almost.

They had one last wish: an election blowout, the kind that would result in a clear victory and send even political junkies on the East Coast to bed by 11:00 p.m. They were out of luck.

While Trump had spent the weeks leading up to November 3 screaming about a "rigged election," pundits and responsible political players had been laying the groundwork for the likelihood that the results would not be known for days. To Trump, the expected rise in the number of mail-in ballots was a source of fraud. To everyone else, it was the result of voting in a global pandemic, at a time when many were still scared to leave their homes.

As the results rolled in that evening, it soon became clear that the possibility of a blowout was over. Trump was winning a host of key states, including Florida. But so was Biden. When Fox News called Arizona for the Democrat, a battleground state, conservatives erupted in rage, setting the stage for a bitter battle. Counting continued well into the night—and, as predicted, for several days. But that didn't stop Trump from addressing the nation from the White House, shortly after 2:00 a.m.

"We were getting ready to win this election," Trump said. "Frankly, we did win this election."

With Trump agitating from the White House, a freshly created Facebook group called "Stop the Steal" was growing by tens of thousands of users per hour, becoming a central part of the infrastructure for wild-eyed claims of brazen election fraud. None of those claims withstood more than a few hours of scrutiny. But that didn't matter. By the time fact-checkers debunked allegations of tampering with electronic voting machines in Pennsylvania, Stop the Steal had already moved on to shouting that poll workers were providing the wrong kind of pen to Republican voters in Arizona that would allow their votes to be invalidated.

The group had grown to more than 360,000 members less than twenty-four hours later when Facebook took it down, citing "extraordinary measures." Pushing false claims of election fraud to a mass audience at a time when armed men were calling for a halt to vote counting outside tabulation centers was an obvious problem, and one that the company knew was only going to get bigger. Stop the Steal had an additional 2.1 million users pending admission to the group when Facebook pulled the plug.

Facebook's leadership would describe Stop the Steal's growth as unprecedented, though Civic staffers could be forgiven for not sharing their sense of surprise. Later analysis would confirm what many immediately suspected. Super-inviters were doing their thing: 30 percent of the Stop the Steal group's members could be traced to just 0.3 percent of users.

There were other signs of trouble. Facebook's classifiers had been good enough to detect that hate speech was spiking nationwide during Stop the Steal's rise, but they weren't good enough to take it down. The "rampant harmful virality" that KX Jin had warned about had arrived, and, as he had warned, Facebook's defenses were too feeble and its rules too convoluted to keep pace. The platform was simply too fast.

On the afternoon of November 5, two days after the end of voting, Facebook "broke glass" on the 2020 U.S. elections. All of a sudden, posts in groups with a history of violating Facebook's rules required admin approval. Users seeking to share election-related content had to click through a notice directing them toward legitimate information sources. Posts classified as having a 70 percent or greater chance of inciting violence began to disappear, and comment threads with abnormally high amounts of hate speech froze. Publishers who scored low on quality metrics received far less distribution, as did posts that were being reshared in long viral chains.

The title of one Break the Glass measure, "Stop Boosting Content from Non-Recommendable Groups," underscored how remedial all this was. Instead of slowing the platform down in the subtle ways that the company's Integrity staff had long recommended, Facebook was pulling the emergency brake. In total, sixty-four separate break-the-glass measures were in place well before the election was called for Biden on November 7.

Facebook's emergency calming measures worked as planned—and were undone as planned, too. Less than a month later, when Facebook metrics for violence and incitement had simmered down to pre-election levels, the company began rolling them back. Strengthened demotions of incitement to violence and distribution penalties for users who repeatedly posted misinformation were among the first to go.

On December 2, roughly two hundred members of Facebook's Civic Integrity team signed on for an all-hands meeting. No one was quite sure what the team would focus on now that the election was past. Everyone knew there was tension between Civic, Public Policy, and a number of Product teams, and some sort of change was in order. It was time, they assumed, for yet another reorg.

Rosen kicked off the meeting by thanking the Civic team, telling them they had made a difference. The work they did was hard, interdisciplinary, and important, and now that the election was behind the company it was time to think more methodically. Civic's efforts didn't scale well—so much of what the team did was ad hoc. Fixing that was going to require closer work with other teams inside the company.

"And so I'm very excited for this next phase in our approach to integrity," he told everyone, before giving the floor to Chakrabarti to talk about what Rosen called "the evolution of the Civic org."

"Cool. All right," Chakrabarti began, sounding less excited than his boss. Civic had been in existence for 1,961 days, he said, recounting how the team first worked on boosting voter registrations before turning its attention to misinformation, tackling wave after wave of crisis both domestic and foreign. The team had done what it could as a stand-alone organization, Chakrabarti said, but it needed the rest of Facebook to succeed. Civic would therefore be dissolved, he announced, and he would be stepping down.

"Over the years, many of you have told me that this team is the first time you felt a true sense of kinship with others who believe in consciously building platforms," he told the audience, acknowledging that breaking up Civic "could feel like a bit of a strange reward."

"As my parting thoughts, what I want you all to see is that the seeds of the Civic values are now within each and every one of you," Chakrabarti said, urging his staff to spread the ethos of being "selfless, constructive, representative, protective, fair, and conscious" across the company.

If Chakrabarti's language sounded more fitting for a missionary than a mid- to senior-level product manager at a tech company, there was a reason. In the politest way possible, he was about to call Facebook's leadership a bunch of heathens.

"When you see a serious problem that we aren't yet taking seriously, you can have the courage to push for expanding our sense of responsibility. When you see us at risk of putting our short-term interests above the long-term needs of the community, you can express yourself constructively and respectfully," he said. "And when we see our platforms further entrenching the powerful rather than democratizing voice, maintain your faith in our company's stated mission."

"That's how you honor the legacy of what we've built here," he concluded. "And in essence, that's how you can do your civic duty."

With that, Chakrabarti turned the meeting back over to Rosen and the other executives who would be inheriting his staff. After a short Q&A, Rosen thanked everyone for their time, and that was that.

Coming in the middle of a pandemic, there was no way for colleagues to commiserate in person. Everyone simply logged off the call. What had been a team was now a couple of hundred ex–Civic staffers, all of them sitting at home with no idea what would become of their work. Among them was Frances Haugen, a thirty-five-year-old product manager who had been at Facebook a little less than a year and a half.

14

I had blind messaged Haugen and a couple dozen of her colleagues at Civic just after the election, reaching out on LinkedIn to say I was familiar with their work, knew there was some internal tension around it, and thought their efforts deserved attention.

In accordance with company rules on media contacts—and the nondisclosure agreement every employee signs upon accepting a job at Facebook—several employees dutifully forwarded my message to Facebook's PR team, and a couple of others sent me polite notes declining to speak. But Haugen had been sitting on it for a month. She had read the stories I had coauthored in the *Journal* about Facebook's willingness to tolerate hate speech and rule violations in India, and how its head of India Policy had openly discussed her disdain for opponents of the country's ruling party.

Haugen had joined Facebook because she wanted to help the company fix its platform. She had deeply personal reasons for feeling that way.

Haugen had grown up in Iowa, the daughter of college professors. She had studied electrical and computer engineering at Olin College, before heading to Silicon Valley in the mid-aughts, a time in which tech was exceptionally kind to twentysomethings with her skill set.

At Google, she worked on the company's efforts to digitize books and make them searchable, a project sufficiently obscure that, even as a junior employee, she ended up with a top role. The work yielded a few patents and an unusually hands-on familiarity with building products to extract and filter information. She did well at it. After a few years, Google paid for her to get an MBA at Harvard. Haugen returned to the company in 2011, ready to resume her career's ascent.

Instead, she nearly died of an autoimmune disorder. Medical records from the time show a private health horror. Nothing was visibly wrong with Haugen, but she went from riding a bicycle as far as a hundred miles a day to struggling to move around without a wheelchair.

Haugen fell behind in her work and resigned at the beginning of 2014. Two months later, she landed in the intensive care unit for three weeks. By the time doctors figured out that the cause was a massive blood clot in her thigh, she had sustained permanent damage to nerves in her hands and feet.

Haugen was a shell of her former self. She couldn't walk up a flight of stairs or hold down a job and was afflicted with excruciating pain from an invisible illness. Living like that indefinitely seemed unthinkable.

A family connection helped bring her back from the edge. Hired to assist Haugen with errands while she was working on her convalescence, the young man became a central support for her after a year spent largely homebound. He went out to get groceries, took her to doctors' appointments, and helped her with physical therapy.

"He was kinda lost, I was lost—we were lost together," Haugen would later tell me. "It was a really important friendship. And then I lost him to being convinced George Soros was running the world's economy."

A Bernie Sanders supporter in 2016, the man had begun spending increasing amounts of time on the online message board 4Chan and the darker corners of Reddit, after the Vermont senator lost the Democratic nomination to Hillary Clinton. His worldview at the time, the man would later tell me, was a mixture of the occult, white nationalism, and the Singularity, the theoretical point at which human consciousness is surpassed by machines.

Haugen's health had improved enough for her to go back to work, first building algorithms to extract data from user-posted photos at Yelp and then joining Pinterest, where she was the lead product manager for one of the platform's central content recommendation systems.

These were good, interesting jobs in a hot field, but Haugen felt like a bit of an outsider. Part of it was that she had fallen behind in her career—she was sometimes recruited for jobs reporting to people she had hired as interns at Google. But her remove went deeper than that. Even after her recovery, the pain and isolation of a long illness set her apart from young peers who had spent their late twenties healthy and social.

Her optimism about social media had been damaged, too. Before she got sick, Haugen had reflexively accepted the idea that social networks were good for the world, or at worst neutral and entertaining. Watching a close friend poison himself on insane, hateful conspiracies blew a hole in that.

"It's one thing to study misinformation. It's another to lose someone to it," she said. "A lot of people who work on these products only see the positive side of things."

Recruiters from Facebook had been trying to poach Haugen for a long time. The regular check-ins weren't so much flattering as a reflection of the fact that Facebook was doubling in size every two years—it needed qualified bodies.

Haugen had never bitten. But when a recruiter reached out at the end of 2018, she said she might be interested if the job touched on disinformation and democracy. As she went through the interview process, she told people that a close friend had been radicalized on the internet and she wanted to help Facebook prevent its users from going down similar paths.

Among her selling points, she noted in a cover letter, was her history of working on sensitive corporate projects "with discretion."

She started at Facebook, working on Civic Integrity, in June 2019.

Haugen was initially asked to build tools to study civic misinformation outside the scope of Facebook's third-party fact-checking. Then, amid fears that Russians were seeding misinformation targeting activist communities and police officers, she was reassigned to building a system that would detect the practice. Her team, consisting of her and four other new hires, was given three months, a schedule she considered implausible. She didn't succeed and got a poor initial review.

"People at Facebook accomplish what needs to be done with far less resources than anyone would think possible," she recalled Chakrabarti telling her. The line was meant to be motivating, but Haugen found it a threadbare excuse on behalf of a company that wasn't investing enough in safety work.

Looking around the larger Integrity team, she saw small bands of employees confronting massive problems. The core team responsible for detecting and combating human exploitation, which included slavery, forced prostitution, and organ selling, included just a few investigators. "I would ask why more people weren't being hired," she recalled. "Facebook acted like it was powerless to staff these teams."

A resource shortage within one of the world's most profitable companies wasn't the only problem she saw. "I was surrounded by smart, conscientious people who every day discovered ways to make Facebook safer," she said. "Unfortunately, safety and growth routinely traded off—and Facebook was unwilling to sacrifice even a fraction of percent of growth."

Haugen came to see herself and the Civic team as an understaffed cleanup crew. Facebook had hired a group of clever and passionate people to fix problems with potentially life-or-death consequences, but it hadn't given them the resources or power to act.

A lot upset Haugen, from Facebook's failure to address human trafficking to the platform's hidden coddling of powerful figures. Above all, she worried that the company was ignoring the dangers its platform posed in poorer countries that were gaining access to the internet for the first time. Myanmar's social media–fueled genocide wasn't a fluke, she believed. It was a template. Facebook was consistently turning up the temperature of societal discourse. If the company didn't change course, its products would kill a lot of people.

Plenty of Haugen's colleagues had reached a similar conclusion. Integrity employees whose projects were blocked sometimes spoke of "stress" and "burnout" before they finally decided to leave the company. Haugen came to believe that the white-collar jargon didn't do their condition justice. She began thinking of Facebook's integrity work as causing moral injury, a term coined by American psychologists to describe a set of PTSD-like symptoms experienced by Vietnam War veterans who had been a party to something that betrayed their deeply held convictions.

Haugen didn't want to feel like that, and she knew she wasn't a big enough fish to accomplish anything by simply badmouthing the company on her way out the door. She was also generally unimpressed with the quality of public discussion around social media companies. Thousands of articles had been written about whether Facebook should remove specific posts, whether it should be broken up, whether it was biased in favor of or against conservatives. But what really mattered, she believed, was Facebook's mechanics. The company had designed its platform in a way that was inherently incendiary, unstable, and prone to manipulation. People outside the company needed to know the details.

On a mid-December afternoon in 2020, I drove to a trailhead at Redwood Regional Park in the Oakland Hills. It was a temperate day, but Haugen emerged from her car wearing a heavy parka, fleece, gloves, leggings, and ankle warmers. As joggers passed by in shorts, Haugen told me the story of her nerve damage. Even a slight chill still meant hours of excruciating pain.

Haugen had texted me on an encrypted messaging app the evening of the all-hands meeting in which Civic's dismantling had been announced.

She had spent the last several months taking notes about her misgivings, as she grew more certain that Facebook wasn't committed to integrity work. Whenever something bothered her or felt significant, she wrote it down. She didn't know what she was going to do with this material. She had reached out to a tech advocacy nonprofit, but her contact there told her it wasn't in the business of whistleblowing.

That Saturday afternoon, we walked for a few minutes, the trail a little too heavily trafficked by joggers and mountain bikers, before pulling off into a clearing to have a real chat. As she ran through her work at the company, she frequently stopped to quiz me on what I knew of various Facebook systems and processes. There was no doubt she was auditioning me: talking to a reporter would be a breach of her nondisclosure agreement with Facebook and potentially the end of her tech career. Beyond asking how I intended to keep her identity safe, she was looking for evidence that talking to me was worth the risk.

We spoke until an early dusk fell below the big trees. On the way back to our cars, Haugen suggested we start meeting regularly. At the very least, she said, answering questions might clarify her thinking and help her identify what was important to document before she left Facebook.

That last bit set Haugen apart. By the time I met her, I had spoken to dozens of former employees who had outlined a general constellation of concerns similar to hers. The details varied—some former employees were incensed about perceived political pandering, others about the dangers of algorithmic content ranking or addictive design. But squint a little and the outlines of a general story emerged. The company had concrete evidence that its products could cause societal-grade harm, but its leaders weren't willing to imperil corporate self-interest by self-correcting.

I knew from experience that knowledgeable employees who reached that conclusion tended to quit, usually quietly. Less frequently they publicly excoriated the company or discussed the details of their work.

But they pretty much never did what Haugen was now contemplating. Rather than smuggle out a stray document or the details of a specific project that had gone awry, Haugen was contemplating spending months inside the company for the specific purpose of investigating Facebook. If Haugen had reached out to me first, rather than the other way around, I would have been paranoid. Though Facebook had never tried to play tricks on me, a well-placed employee dangling such extraordinary assistance at a first meeting seemed too good to be true.

I spent half my drive home trying to imagine how she could be the bait in some elaborate setup and the other half readying myself for disappointment when she didn't follow through. Taking stock of everything that could go wrong, I also found myself a little uncomfortable with the intensity of her convictions. I hadn't betrayed my skepticism during our meeting, but Haugen had twice in our talk said that she was motivated to help me because, if she didn't expose what was known inside Facebook, millions of people would likely die.

With the United Nations having already blamed the company for driving ethnic cleansing in Myanmar, her claim couldn't be completely dismissed out of hand. But it still seemed grandiose.

After two years of covering the company, I understood that its products, and those of its social media competitors, could alter politics and the flow of information in ways that could shape and sometimes end lives. But the assertion that smartphone apps, even popular ones, had the destructive force she claimed still seemed hyperbolic.

The role of Facebook and other platforms in events that would come just a few weeks later left me a little less sure.

15

They described one another as grifters, prima donnas, and clowns. They lied, reflexively and clumsily, in pursuit of money and relevance. Through the power of social media they would change the course of American political history.

In the aftermath of January 6, reporters and investigators would focus on intelligence failures, White House intrigue, and well-organized columns of white nationalists. Those things were all real. But a fourth factor came into play: influencers.

Trump was, in a way, the ultimate right-wing influencer, skilled at gaming his platform of choice, Twitter, to bend the news cycle to his will. Behind him came a caravan of crazies, hoping to influence their own way to stardom. For many of them, the platform of choice was Facebook.

The Stop the Steal Facebook group was born out of a mid-Election Day chat between a mother-daughter duo named Amy and Kylie Kremer, conservative activists with a history of feuding with rival Tea Party groups over fundraising. Formed with a few clicks by Kylie, the group's name was chosen in an attempt to squat on the #StopTheSteal hashtag already trending on Twitter and Facebook.

In later interviews with the special January 6 Committee set up by the House to investigate the events of that fateful day, neither woman could explain why their group took off compared to other similarly named entities. The fact that the Kremers created it via an already-sizable Facebook page with verified status couldn't have hurt. Within a few hours of its creation, Stop the Steal's membership had expanded into the hundreds of thousands, with both everyday users and vocal online activists flocking to it like moths to a porch light.

"I think it was growing by 1,000 people every 10 seconds, which kind of broke a lot of algorithms," Kylie Kremer told investigators.

Facebook took it down, but that didn't mean the movement, and its attendant influencers, stopped gaining ground. Things got messy almost immediately, as personalities vying for online attention through outrageous words and behavior clashed. Ali Alexander, an ex-felon and far-right conspiracy theorist, set up a fundraising website and an LLC with the "Stop the Steal" name, soliciting donations that the Kremers and other activists would later accuse him of pocketing. Brandon Straka, a New York City hairstylist and founder of #WalkAway, a movement that had grown to more than a million followers on Facebook to ostensibly encourage Democrats to leave their party, also joined in.

The Kremers' first attempt at convening the influencers on a conference call devolved into shouting. Straka sparred with Kylie Kremer, later telling investigators, "I found her to be emotionally unstable, and a—and incompetent." Kylie Kremer in turn clashed with the conspiracy theorist media personality Alex Jones, who was also trying to get in on the action. She filed a report with DC police accusing him of "threats to bodily harm" after he allegedly threatened to push her off the stage while reportedly yelling, "I'm gonna do it. I'm gonna do it. I'm going to take over."

Things ramped up once Trump tweeted, on December 19, that there would be a "big protest in D.C. on January 6th. Be there, will be wild!" The tweet blindsided the Kremers and White House staff alike, and it didn't unite anyone so much as up the ante of the squabbling. The stakes were getting bigger, and anyone who was anyone in the world of right-wing influencers wanted a piece of the action.

The seventy-three-year-old heiress to the Publix supermarket fortune donated $3 million in total to the effort, some of it going to Jones, some to Trump adviser Roger Stone, but a huge chunk—$1.25 million—going to Charlie Kirk, the founder of the conservative student group Turning Point USA, to "deploy social media influencers to Washington" and "educate millions." When investigators later confronted Kirk with documents showing he had billed the heiress for $600,000 of buses that were never chartered, he responded by invoking his Fifth Amendment right against self-incrimination.

This band of self-declared patriots came together at a unique moment in American history, with social media coming fully into its power, and they had a ready audience in the social media–obsessed president in the White House. Trump wanted the likes of Alexander and Jones to speak at his January 6 event, according to texts sent by one of his aides, Katrina Pierson. Or as she put it in a text to Kylie Kremer: "He likes the crazies." There was a nominal division between the "crazies" on one side and the Kremers on the other, but all were coming together to make sure January 6 would be unforgettable.

"I mean, there were so many things that were being said or pushed out via social media that were just concerning," Kylie Kremer told investigators, while defending the decision to maintain a loose alliance with what she variously called mercenary, larcenous, and quite possibly mentally ill social media activists who were posting about civil war, 1776, and their willingness to die for liberty. "It took all of us getting the messaging out to get all the people that came to DC," she said.

The influencers' belligerence was the source of their power. "The more aggressive people, like the Alis and all those guys, they began to get a little bit more prominence because of the language that they were using," Pierson told investigators.

Trump may have promoted the Kremers' official January 6 protest on his Twitter account, but, in the end, one activist noted, they collected only 20,000 RSVPs on Facebook. Ali's bootleg site, pumped with louder language and even wilder conspiracy theories, pulled in 500,000.

Pierson's "crazies" were, in fact, the luminaries of Zuckerberg's Fifth Estate.

"These people had limited abilities to influence real-life outcomes—if Ali Alexander had put out a call for people to march on the Capitol, a few dozen people would have shown up," said Jared

Holt, who researched the run-up to January 6 for the Facebook-funded Atlantic Council's Digital Forensic Research Lab. "But it's the network effects where they took hold, where people who are more respectable and popular than Ali reshape his content."

To keep the influencers hyping the January 6 rally—but nowhere near the president himself—Pierson helped broker a deal for what she called "the psycho list" to speak at a different event on January 5. Amid a frigid winter drizzle in DC's Freedom Plaza, Ali and Straka ranted alongside Jones, disgraced former New York police commissioner Bernard Kerik, and the guy behind the "DC Draino" meme account, which had 2.3 million followers on Instagram alone.

The next day, at the real rally, the Kremers instructed security to be ready if Ali, Jones, or Straka attempted to rush the stage and seize the microphone by force.

Straka told investigators that he would have liked to speak on January 6 himself, but, barring that, he made the best of things. "I've got my camera, I've got my microphone," he recalled thinking. "I am going to turn it into an opportunity to create content for my audience."

The story of how Facebook's defenses were so soundly defeated by such a hapless crew begins right after the Stop the Steal group's takedown, and it starts right at the top.

Although Facebook had vaguely alleged that it had taken down the group because of prohibited content, the truth was that the group hadn't violated Facebook's rules against incitement to violence, and the platform had no policy forbidding false claims of election fraud. Based on the group's obvious malignancy, however, Bickert and Facebook's Content Policy team had declared a "spirit of the policy" violation, a rare but not unheard-of designation that boiled down to "because we say so."

Zuckerberg had accepted the deletion under emergency circumstances, but he didn't want the Stop the Steal group's removal to become a precedent for a backdoor ban on false election claims. During the run-up to Election Day, Facebook had removed only lies about the actual voting process—stuff like "Democrats vote on Wednesday" and "People with outstanding parking tickets can't go to the polls." Noting the thin distinction between the claim that votes wouldn't be counted and that they wouldn't be counted accurately, Chakrabarti had pushed to take at least some action against baseless election fraud claims.

Civic hadn't won that fight, but with the Stop the Steal group spawning dozens of similarly named copycats—some of which also accrued six-figure memberships—the threat of further organized election delegitimization efforts was obvious.

Barred from shutting down the new entities, Civic assigned staff to at least study them. Staff also began tracking top delegitimization posts, which were earning tens of millions of views, for what one document described as "situational awareness." A later analysis found that as much as 70 percent of Stop the Steal content was coming from known "low news ecosystem quality" pages, the commercially driven publishers that Facebook's News Feed integrity staffers had been trying to fight for years.

Civic had prominent allies in this push for intelligence gathering about these groups, if not for their outright removal. Facebook had officially banned QAnon conspiracy networks and militia groups earlier in the year, and Brian Fishman, Facebook's counterterrorism chief, pointed to data showing that Stop the Steal was being heavily driven by the same users enthralled by fantasies of violent insurrection.

"They stood up next to folks that we knew had a track record of violence," Fishman later explained of Stop the Steal.

But Zuckerberg overruled both Facebook's Civic team and its head of counterterrorism. Shortly after the Associated Press called the presidential election for Joe Biden on November 7—the traditional marker for the race being definitively over—Molly Cutler assembled roughly fifteen executives that had been responsible for the company's election preparation. Citing orders from Zuckerberg, she said the election delegitimization monitoring was to immediately stop.

Though Zuckerberg wasn't there to share his reasoning, Rosen hadn't shied away from telling Chakrabarti that he agreed with Zuckerberg's decision—an explanation that Chakrabarti found notable enough to make a record of. He quoted Rosen in a note to the company's HR department as having told him that monitoring efforts to stop the presidential transition would "'just create momentum and expectation for action' that he did not support."

The sense that the company could put the election behind it wasn't confined to management. Ryan Beiermeister, whose work leading the 2020 Groups Task Force was widely admired within both Civic and the upper ranks of Facebook's Integrity division, wrote a note memorializing the strategies her team had used to clean up what she called a "powderkeg risk."

Beiermeister, a recent arrival to Facebook from the data analysis giant Palantir, congratulated her team for the "heroic" efforts they made to get Facebook's senior leadership to sign off on the takedowns of toxic groups. "I truly believe the Group Task Force made the election safer and prevented possible instances of real world violence," she concluded, congratulating the team's thirty members for the "transformative impact they had on the Groups ecosystem for this election and beyond."

Now, with the election crisis seemingly over, Facebook was returning its focus to engagement. The growth-limiting Break the Glass measures were going to have to go.

On November 30, Facebook lifted all demotions of content that delegitimized the election results. On December 1, the platform restored misinformation-rich news sources to its "Pages You Might Like" recommendations and lifted a virality circuit breaker. It relaxed its suppression of content that promoted violence the day after that, and resumed "Feed boosts for non-recommendable Groups content" on December 7. By December 16, Facebook had removed the caps on the bulk group invitations that had driven Stop the Steal's growth.

Only later would the company discover that more than four hundred groups posting pro-insurrectionist content and false claims of a stolen election were already operating on Facebook when the company lifted its restrictions on bulk invitations. "Almost all of the fastest growing FB Groups were Stop the Steal during the period of their peak growth," the document noted.

A later examination of the social media habits of people arrested for their actions on January 6 found that many "consumed fringe Facebook content extensively," much of it coming via their membership in what were sometimes hundreds of political Facebook groups. On average, those groups were posting twenty-three times a day about civil war or revolution.

Facebook had lowered its defenses in both the metaphorical and technical sense. But not all the degradation of the company's integrity protections was intentional. On December 17, a data scientist flagged that a system responsible for either deleting or restricting high-profile posts that violated Facebook's rules had stopped doing so. Colleagues ignored it, assuming that the problem was just a "logging issue"—meaning the system still worked, it just wasn't recording its actions. On the list of Facebook's engineering priorities, fixing that didn't rate.

In fact, the system truly had failed, in early November. Between then and when engineers realized their error in mid-January, the system had given a pass to 3,100 highly viral posts that should have been deleted or labeled "disturbing."

Glitches like that happened all the time at Facebook. Unfortunately, this one produced an additional 8 billion "regrettable" views globally, instances in which Facebook had shown users content that it knew was trouble. The company would later say that only a small minority of the 8 billion "regrettable" content views touched on American politics, and that the mistake was immaterial to subsequent events. A later review of Facebook's post-election work tartly described the flub as a "lowlight" of the platform's 2020 election performance, though the company disputes that it had a meaningful impact. At least 7 billion of the bad content views were international, the company says, and of the American material only a portion dealt with politics. Overall, a spokeswoman said, the company remains proud of its pre- and post-election safety work.

Facebook had never gotten out of the red zone on Civic's chart of election threats. Now, six weeks after the election, the team's staffers were scattered, Chakrabarti was out, and protections against viral growth risks had been rolled back.

In the days leading up to January 6, the familiar gauges of trouble—hate speech, inflammatory content, and fact-checked misinformation—were again ticking up. Why wasn't hard to guess. Control of the Senate depended on a Georgia runoff election scheduled for January 5 and Trump supporters were beginning to gather in Washington, DC, for the protest that Trump had promised would "be wild!"

The Counterterrorism team reporting to Brian Fishman was tracking pro-insurrection activity that he considered "really concerning." By January 5, Facebook was preparing a new crisis coordination team, just in case, but nobody at the company—or anywhere in the country, really—was quite ready for what happened next.

On January 6, speaking to a crowd of rowdy supporters, Trump again repeated his claim that he had won the election. And then he directed them toward the Capitol, declaring that, "If you don't fight like hell, you're not going to have a country anymore." Floods of people streamed toward the Capitol and, by 1:00 p.m., rioters had broken through the outer barriers around the building.

Fishman, out taking a walk at the time, sprinted home, according to a later interview with the January 6 Committee. It was time to start flipping those switches again. But restoring the safeguards that Facebook had eliminated just a month earlier came too late to keep the peace at Facebook, or anywhere else. Integrity dashboards reflected the country's social fabric rending in real time, with reports of false news quadrupling and calls for violence up tenfold since the morning. On Instagram, views of content from what Facebook called "zero trust" countries were up sharply, suggesting hostile entities overseas were jumping into the fray in an effort to stir up additional strife.

Temperatures were rising on Workplace, too. For those on the front lines of the company's response, the initial silence from Facebook's leadership was deafening.

"Hang in there everyone," wrote Mike Schroepfer, the chief

technology officer, saying company leaders were working out how to "allow for peaceful discussion and organizing but not calls for violence."

"All due respect, but haven't we had enough time to figure out how to manage discourse without enabling violence?" an employee snapped back, one of many unhappy responses that together drew hundreds of likes from irate colleagues. "We've been fueling this fire for a long time and we shouldn't be surprised that it's now out of control."

Shortly after 2:00 p.m., rioters entered the Capitol. By 2:20 p.m., the building was in lockdown.

Several hours passed before Facebook's leadership took their first public steps, removing two of Trump's posts. Privately, the company revisited its determination that Washington, DC, was at "temporarily heightened risk of political violence." Now the geographic area at risk was the entire United States.

As rioters entered the Senate chamber and offices around the building, while members of Congress donned gas masks and hid where they could, Facebook kept tweaking the platform in ways that might calm things down, going well past the set of Break the Glass interventions that it had rolled out in November. Along with additional measures to slow virality, the company ceased auto-deleting the slur "white trash," which was being used quite a bit as photos of colorfully dressed insurrectionists roaming the Capitol went viral. Facebook had bigger fish to fry than defending rioters from reverse racism.

Enforcement operations teams were given a freer hand, too, but it wasn't enough. Everything was going to have to be put on the table, including the near-inviolability of Trump's right to use the platform.

As evening descended on DC, Trump released a video on the advice of several advisers, who pitched it as an attempt to calm tensions. "We have to have peace, so go home," the embattled president said. But he couched it in further declarations that the election had been stolen and also addressed the rioters, saying, "We love you. You're very special." Facebook joined YouTube and Twitter in taking it down, and then suspended his account for twenty-four hours. (It

would go on to extend the ban through Biden's inauguration, scheduled for January 20, before deciding to boot him from the platform
indefinitely.)

Zuckerberg remained silent through January 6, leaving Schroepfer to calm tensions the following morning. "It's worth stepping back
and remembering that this is truly unprecedented," he wrote on
Workplace. "Not sure I know the exact right set of answers but we
have been changing and adapting every day—including yesterday."

More would yet be needed. While the company was restricting
its platform in ways it had never attempted in a developed market,
it wasn't enough to suppress "Stop the Steal," a phrase and concept
that continued to surge. The Public Policy team, backed by Facebook's leadership, had long held that the company not remove, or
even downrank, content unless it was highly confident that it violated the rules. That worked to the advantage of those deploying
#StopTheSteal. Having ruled that the claims of a stolen election
weren't inherently harmful, Facebook had left the groups as a whole
alone after taking down the Kremers' original group.

"We were not able to act on simple objects like posts and comments because they individually tended not to violate, even if they
were surrounded by hate, violence and misinformation," a later
report found.

Only after the fires were out in the Capitol, and five people were
dead, did Facebook realize how badly it had gotten played. A core
group of coordinated extremists had hyperactively posted, commented, and reshared their movement into existence.

Stop the Steal wasn't just another hashtag. It was a "rallying point
around which a movement of violent election delegitimization
could coalesce," a later review said. It also gently suggested who at
the company was to blame: leadership and Public Policy. "Seams" in
platform rules had allowed "the larger wave of the movement seep-
ing through the cracks," the review found.

There was no time to point fingers at the C-suite, or anywhere else
for that matter. #StopTheSteal was surging in the wake of January 6.
No sooner had Integrity teams nuked the hashtag and mapped out
networks of advocates using it than they identified a new threat: the

same insurrectionist community was uniting to take another shot. The new rallying point was the "Patriot Party," which pitched itself as a far-right, Trump-supporting alternative to the Republican Party.

In a gift to Facebook investigators racing to track them, those organizing the new "party" gathered in private, admin-only groups on the platform, essentially providing a roadmap to their central leadership. What Facebook would do with that information, however, was an open question.

Under normal circumstances, Facebook investigators would spend weeks compiling a dossier on the misbehavior of each individual entity it wanted to shut down—and then try to get a mass takedown approved by Facebook's Public Policy team. This time, however, nobody in leadership was worried about overenforcement.

Facebook's new crisis-time approach was the tried-and-true method that bartenders reserve for rowdy drunks: stop serving them and throw the bastards out. The company began heavily downranking the term "Patriot Party" across Facebook and Instagram and started summarily deleting the central nodes of the network promoting it. Ali Alexander's accounts were dead meat. So was Brandon Straka's "Walk Away" movement, as well as those of numerous other Stop the Steal influencers.

The combat between Facebook and a movement it helped birth lasted for weeks. When the company's internal state of emergency was finally lowered from its highest level on January 22, two days after Biden's inauguration, there was no question that the tactics had succeeded. With the flip of a few switches, the megaphone the platform had given the insurrectionists was ripped from their hands.

"We were able to nip terms like 'Patriot Party' in the bud," the later review noted.

Blocking the insurrectionists' attempts to regroup was an accomplishment, though one tempered with apprehension for many former Civic staffers. On leadership's orders, they had just smothered an attempt to organize what might have become a new political party. Ironically, Zuckerberg had earlier rejected much less heavy-handed efforts to make the platform more stable on the grounds that they restricted user voice.

In the immediate aftermath of January 6, Facebook kept its head down. When company sources and Facebook Communications staff returned reporters' phone calls or answered emails, which was rare, they acknowledged a good deal of soul-searching. The closest thing to a defense of the company's performance that anyone offered up was noting that the principal responsibility for the assault on the Capitol belonged to Trump.

On January 7, as most of the world was trying to figure out what the hell had just happened, a brief essay titled "Demand Side Problems" appeared on a blog where Andrew Bosworth posted his thoughts about philosophy and leadership. His thesis was that Facebook's users had the same insatiable desire for hate as Americans had for narcotics. Therefore, Facebook's efforts to suppress hate, while well intentioned, would fail just like the "war on drugs" had—at least "until we make more social progress as a society."

Nobody took notice—this was a Facebook executive's personal philosophy blog, after all—but the post still made quite a statement. A day after a hate- and misinformation-driven riot had shaken democracy, one of Zuckerberg's top lieutenants was blaming the wickedness of society at large—in other words, Facebook's users.

The note was an abridged version of a previous Workplace post Bosworth had made, kicking off a debate with fellow employees in which he'd gone further. Cracking down too hard, he wrote, was a bad idea because hate-seeking users would "just satisfy their demand elsewhere."

Bosworth's argument contained some nuance: As the entity that oversaw the supply of content, Facebook was obligated "to invest huge amounts" in user safety. And the executive—who oversaw Facebook's hardware business, not content moderation—said in other notes that he favored more work to address recommendations' bias toward outrage- and virality-related problems.

But the takeaway from many of the Integrity staffers who read the post was simpler. If people had to be bigots, the company would prefer they be bigots on Facebook.

—

By the time Facebook's Capitol riot postmortems got underway, Haugen and I had been meeting roughly every week in the backyard of my home in Oakland, California. To thwart potential surveillance to the greatest degree possible, I had paid cash for a cheap Samsung phone and given it to Haugen, who I now referred to as "Sean McCabe." The name was her choice, borrowed from a friend in the local San Francisco arts scene who had died just after we met. McCabe had been a troublemaker with a taste for excitement, and she was pretty sure he would have approved of what she was doing.

"Sean" was now using the burner phone to take screenshots of files on her work laptop, transferring the images to a computer that she'd bought in order to have a device that had never been connected to the internet.

One of those files was the resulting document of Facebook's postmortem, titled "Stop the Steal and the Patriot Party: The Growth and Mitigation of an Adversarial Harmful Movement." Many of the people involved in drafting it were former members of Civic. It was illustrated with a cartoon of a dog in a firefighter hat in front of a burning Capitol building. (The hot document would be published by BuzzFeed in April.)

"What do we do when a movement is authentic, coordinated through grassroots or authentic means, but is inherently harmful and violates the spirit of our policy?" the report asked. "We're building tools and protocols and having policy discussions to help us do better next time."

Less than a year and a half earlier, at Georgetown, Zuckerberg had scolded unnamed social media critics who wanted Facebook to "impose tolerance top-down." The journey toward progress, he declared, "requires confronting ideas that challenge us." Facebook was now confronting the possibility that its platforms weren't destined to produce healthy outcomes after all. A small group of hyperactive users had harnessed Facebook's own growth-boosting tools to achieve vast distribution of incendiary content.

In the wake of January 6, Haugen had been brought into a working group meant to address "Adversarial Harmful Networks." It was tasked with identifying the lessons that could be learned from Facebook's already completed Patriot Party work, which was seen as a template for future actions. After mapping out malignant movements and identifying their leadership structures, Facebook could, as needed, undertake a lightning strike to knock them out, then launch mop-up operations to prevent new leaders from regrouping. She and I jokingly branded this new strategy the "SEAL Team Six approach."

The team's official name would undergo many changes, a reflection perhaps of how uncomfortable Facebook was with tackling the toughest issues. "Adversarial Harmful Networks" became "Harmful Topic Communities," "Non-Recommendable Conspiracy Theories," "Non-Violating Harmful Narratives," and—as of the time of writing—"Coordinated Social Harm."

No matter what the team was called, the strategy was the same. In the wake of Stop the Steal and the Patriot Party, the company had begun pivoting its vast data analysis tools to understanding the inner workings of movements that looked like trouble. That cute corgi pup meant business.

After gathering the behavioral data and activities of 700,000 supporters of Stop the Steal, Facebook mapped out the connections among them and began dividing them into ringleaders (those who created content and strategy), amplifiers (prominent accounts that spread those messages), bridgers (activists with a foot in multiple communities, such as anti-vax and QAnon), and finally "susceptible users" (those whose social circles seemed to be "gateways" to radicalism).

Together, this collection of users added up to an "information corridor." Messages that originated among a movement's elite users pulsed through paths connecting Facebook's users and products, with every reshare, reply, and reaction spreading them to an ever-widening audience. Over time, those hidden vectors had become well-worn, almost reflexive, capable of transmitting increasingly bizarre stuff on an unprecedented scale.

Analysis showed that the spread of hate, violence, and misinformation occurred at significantly higher rates along the Stop the Steal information corridor than on the platform at large. But Facebook wasn't trying to filter out the bad stuff anymore. Instead, it was looking to identify and jam the transmission lines.

The work involved a lot of statistics and machine learning, but the core principles were not hard to grasp. To kill a movement, the ringleader accounts should be killed all at once, depriving the movement of its brain. Their lieutenants—who would likely try to replace their leaders in a "backlash" against the removals—could be slapped with strict limits on creating new pages, groups, and posts. Amplifiers and bridgers could merely be de-amplified through downranking. And, finally, the company would seek to prevent connections from forming between "susceptible" people, with Facebook's recommendations actively steering them away from content and users that might take them deeper down whichever rabbit hole they were peering into.

The team ran experiments with these proposals, using a model based on the historical Stop the Steal data. The results suggested that the approach would have worked, kneecapping the movement on Facebook long before January 6. "Information Corridors could have helped us identify the social movement around delegitimization from an array of individually noisy text signals," said one memo, titled "Information Corridors: A Brief Introduction." Other documents that Haugen shared with me also seemed to suggest that Facebook was convinced it could have headed off "the growth of the election delegitimizing movements that grew, spread conspiracy, and helped incite the Capitol Insurrection."

The enthusiasm for the approach was great enough that Facebook created a Disaggregating Networks Task Force to take it further. The company had gone from allowing conspiracy-minded groups of all stripes to flourish, to extreme concern over how people were getting sucked in. In one coordinating call, which Haugen participated in, the head of the task force noted that the company had twelve different teams working on methods to not just break up the leadership of harmful movements but inoculate potential followers against them. "They are vulnerable and we need to protect them," the head of the project declared on the call about "susceptible" audiences.

This concerned Haugen, a progressive Democrat who also had libertarian leanings, the kind of tech employee who made a habit of attending Burning Man. Facebook, she realized, had moved from targeting dangerous actors to targeting dangerous ideas, building systems that could quietly smother a movement in its infancy. She heard echoes of George Orwell's thought police. To her, this was getting creepy, and unnecessary.

Facebook had years of research showing how it could have changed its platform to make it vastly less useful as an incubator for communities built around violent rhetoric, conspiracy theories, and misinformation. It could avoid killing exponentially growing conspiracy groups if it prevented their members from inviting a thousand strangers to join them in a day. It wouldn't need to worry so much about "information corridors" resharing misinformation endlessly if it capped reshares, as Civic had long pushed it to do.

But, following a familiar script, the company was unwilling to do anything that would slow down the platform—so it was embarking on a strategy of simply denying virality to hand-picked entities that it feared. And the work was moving ahead at a high speed.

By the spring of 2021, Facebook was experimenting with shutting down "information corridors" of its "harmful topic communities." It chose as a target Querdenken, a German movement that pushed a conspiracy theory that a Deep State elite, in concert with "the Jews," was pushing COVID restrictions on an unwitting population. Adherents of the movement had attacked and injured police at anti-lockdown protests, but the largest Querdenken page on Facebook still had only 26,000 followers. In other words, Querdenken was small, violent, and short of friends in the German government. That made it an excellent guinea pig.

Facebook wasn't that concerned about killing Querdenken. It just wanted to make sure that it could. As it prepared to run the experiment, Facebook divided the movement's adherents into treatment and control groups, altering their News Feeds accordingly. The plan was to start the work in mid-May.

As it turned out, the company wouldn't have to wait that long for its next chance to use its new tools.

—

Zuckerberg vehemently disagreed with people who said that the COVID vaccine was unsafe, but he supported their right to say it, including on Facebook. That had been the CEO's position since before a vaccine even existed; it was a part of his core philosophy. Back in 2018, the CEO had gone so far as to say the platform shouldn't take down content that denied the Holocaust because not everyone who posted Holocaust denialism "intended" to. (He later clarified to say he found Holocaust denial "deeply offensive," and the way he handled the issue angered Sheryl Sandberg, a fellow Jew, who eventually succeeded in persuading him to reverse himself.)

Under Facebook's policy, health misinformation about COVID was to be removed only if it posed an imminent risk of harm, such as a post telling infected people to drink bleach. "I think that if someone is pointing out a case where a vaccine caused harm, or that they're worried about it, that's a difficult thing to say, from my perspective, that you shouldn't be allowed to express at all," Mr. Zuckerberg had told *Axios* in a September 2020 interview.

But, early in February 2021, Facebook began to realize that the problem wasn't that vaccine skeptics were speaking their mind on Facebook. It was how often they were doing it.

A researcher randomly sampled English-language comments containing phrases related to COVID and vaccines. A full two-thirds were anti-vax. The researcher's memo compared that figure to public polling showing the prevalence of anti-vaccine sentiment in the U.S.—it was a full 40 points lower.

Additional research found that a small number of "big whales" was behind a large portion of all anti-vaccine content on the platform. Of 150,000 posters in Facebook groups that were eventually disabled for COVID misinformation, just 5 percent were producing half of all posts. And just 1,400 users were responsible for inviting half of all members. "We found, like many problems at FB, this is a head-heavy problem with a relatively few number of actors creating a large percentage of the content and growth," Facebook researchers would later note.

One of the anti-vax brigade's favored tactics was to piggyback on posts from entities like UNICEF and the World Health Organization encouraging vaccination, which Facebook was promoting free of charge. Anti-vax activists would respond with misinformation or derision in the comments section of these posts, then boost one another's hostile comments toward the top slot with almost incomprehensible zeal. Some were nearing Facebook's limits on commenting, which was set at three hundred times per hour. As a result, English-speaking users were encountering vaccine skepticism 775 million times each day.

As with previous malign efforts, such as Russian trolls or Stop the Steal, it was hard to gauge how effective these tactics were in persuading people to avoid the vaccine. But directionally the effects were clear. People logging onto Facebook's platforms would log off believing, at the very least, that the vaccine was more controversial than it actually was.

Investigation of the movement uncovered no evidence of inauthentic behavior or disallowed tactics. That meant it was again time for information corridor work. The team created to fight "Dedicated Vaccine Discouragement Entities" set the goal of limiting the anti-vax activity of the top 0.001 percent of users—a group that turned out to have a meaningful effect on overall discourse.

By early May 2021, as it was nearing the time to launch its Querdenken experiment, which would end up running for a few months, the situation with COVID misinformation had gotten so bad that the company found itself dipping into its Break the Glass measures. As recently as six months earlier, Facebook had hoped it would never need to use those measures in the United States at all. Now it was deploying them for the third time in half a year.

Unlike earlier conflagrations, Facebook couldn't blame this round on Trump. Since the fall of 2016, the company had, not unreasonably, pointed to the erratic president as the precipitating factor behind fake news, racial division, and election delegitimization. He may have unleashed a new vitriol in American politics, but now he was out of office and off the platform. This movement had its own set of originators.

A state of crisis was becoming the norm for Facebook, and the Integrity team's approach to its work was beginning to reflect that. Facebook started working on building a "kill switch" for each of its recommendation systems. Integrity team leaders began to espouse a strategy of "Always-On Product Iteration," in which every new scramble to contain an escalating crisis would be incorporated into the company's plans for the next catastrophe.

"Yay for things incubating on Covid and becoming part of our general defense," a team leader wrote, putting a positive spin on expectations for an increasingly unstable world.

Even as Facebook prepared for virally driven crises to become routine, the company's leadership was becoming increasingly comfortable absolving its products of responsibility for feeding them. By the spring of 2021, it wasn't just Boz arguing that January 6 was someone else's problem. Sandberg suggested that January 6 was "largely organized on platforms that don't have our abilities to stop hate." Zuckerberg told Congress that they need not cast blame beyond Trump and the rioters themselves. "The country is deeply divided right now and that is not something that tech alone can fix," he said.

In some instances, the company appears to have publicly cited research in what its own staff had warned were inappropriate ways. A June 2020 review of both internal and external research had warned that the company should avoid arguing that higher rates of polarization among the elderly—the demographic that used social media least—was proof that Facebook wasn't causing polarization.

Though the argument was favorable to Facebook, researchers wrote, Nick Clegg should avoid citing it in an upcoming opinion piece because "internal research points to an opposite conclusion." Facebook, it turned out, fed false information to senior citizens at such a massive rate that they consumed far more of it despite spending less time on the platform. Rather than vindicating Facebook, the researchers wrote, "the stronger growth of polarization for older users may be driven in part by Facebook use."

All the researchers wanted was for executives to avoid parroting a claim that Facebook knew to be wrong, but they didn't get their wish. The company says the argument never reached Clegg. When he published a March 31, 2021, Medium essay titled "You and the Algorithm: It Takes Two to Tango," he cited the internally debunked claim among the "credible recent studies" disproving that "we have simply been manipulated by machines all along." (The company would later say that the appropriate takeaway from Clegg's essay on polarization was that "research on the topic is mixed.")

Such bad-faith arguments sat poorly with researchers who had worked on polarization and analyses of Stop the Steal, but Clegg was a former politician hired to defend Facebook, after all. The real shock came from an internally published research review written by Chris Cox.

Titled "What We Know About Polarization," the April 2021 Workplace memo noted that the subject remained "an albatross public narrative," with Facebook accused of "driving societies into contexts where they can't trust each other, can't share common ground, can't have conversations about issues, and can't share a common view on reality."

But Cox and his coauthor, Facebook Research head Pratiti Raychoudhury, were happy to report that a thorough review of the available evidence showed that this "media narrative" was unfounded. The evidence that social media played a contributing role in polarization, they wrote, was "mixed at best." Though Facebook likely wasn't at fault, Cox and Raychoudhury wrote, the company was still trying to help, in part by encouraging people to join Facebook groups. "We believe that groups are on balance a positive, depolarizing force," the review stated.

The writeup was remarkable for its choice of sources. Cox's note cited stories by *New York Times* columnists David Brooks and Ezra Klein alongside early publicly released Facebook research that the company's own staff had concluded was no longer accurate. At the same time, it omitted the company's past conclusions, affirmed in another literature review just ten months before, that Facebook's recommendation systems encouraged bombastic rhetoric from publishers and politicians, as well as previous work finding that seeing vicious posts made users report "more anger towards people with different social, political, or cultural beliefs." While nobody could reliably say how Facebook altered users' off-platform behavior, how the company shaped their social media activity was accepted fact. "The more misinformation a person is exposed to on Instagram the more trust they have in the information they see on Instagram," company researchers had concluded in late 2020.

In a statement, the company called the presentation "comprehensive" and noted that partisan divisions in society arose "long before platforms like Facebook even existed." For staffers that Cox had once assigned to work on addressing known problems of polarization, his note was a punch to the gut. Their patron—someone who had read their own far more rigorous research reviews, been briefed on analyses and experiments, and championed their plans to address the design flaws of groups—was saying that the problem they were assigned to was as real a threat as werewolf attacks.

"We had all celebrated when Cox came back. Before he left he had been a counterweight in the org, someone who'd say, 'This is not something I think we should be doing,'" recalled one director, who cited Cox as his inspiration for standing up to the company's leadership in ways that damaged his career. "I may have gotten my estimation of the man wrong."

16

Haugen began building a stable of internal documents, though she had sound security reasons not to explore too widely within Facebook's systems. She was part of an Investigations and Intelligence team, which gave her ready access to some extremely sensitive stuff. If someone asked why she was snooping around the "information corridor" work, she would have a readier answer than an employee in, say, Instagram's ad sales department. Still, she knew that Facebook logged what people viewed and did on Workplace. We discussed the information she was finding, but the goal wasn't to rush things into print—neither of us was eager to kick off a leak investigation. We jokingly adopted Rosen's motto for success—"Good Work, Consistently, Over a Long Period of Time"—as inspiration for Haugen's new side gig foraging for internal Facebook information.

Nevertheless, there was no question that, by the spring of 2021, Haugen was starting to drift a little. Our outdoor backyard lunches of takeout sushi, where she would unpack her experience at Facebook and thoughts about the platform's design, could be revelatory, but they were becoming fewer and farther between. Sometimes as much as a week or two would go by before she would respond to a Signal message. When we did meet, she began talking about potentially moving to the Sierra Nevada foothills or quitting Facebook to found an open-source data analysis company. She was taking an advisory role in the launch of a cryptocurrency startup being pursued by friends.

Haugen later acknowledged that the strain of being a self-appointed mole was starting to get to her. While COVID-era remote work made exfiltrating information much easier—Haugen couldn't have gotten away with snapping phone pics of sensitive documents at the office—she was still under a lot of stress. How could her team, which included intelligence analysts drawn from the U.S. government, not see through her?

As much as Haugen believed that the majority of her former Civic colleagues shared her disillusionment with Facebook, she didn't have their permission to make their work public. Past leaks—namely those by the right-wing undercover provocateurs of "Project Veritas"—had produced death threats and the deployment of round-the-clock corporate security details for data scientists. I had assured her that the *Journal* would keep non-executives' names out of print, but there was no question that airing her colleagues' work in the *Journal* would cause people trouble.

The initial burst of energy she'd derived from working with a journalist was long since gone, and even thinking about what to do with the information was stressful. We still talked over her plans to do something really big, but her information gathering seemed to be slowing down. A principal reason for our in-person meetings, she said, was that she needed the structure of what she began calling "study hall."

"Figuring what information would bring about meaningful change is a hard thing," she said later. "And the idea that I couldn't be 100 percent transparent with the people in my life was hard on me. It really wore me down."

By March, I was routinely failing to pin Haugen down for meetings. Given that she had already spent dozens of hours walking me through Facebook's internal recommendation and enforcement systems, I was hardly in a position to whine, but the lengthening silences stung nonetheless.

"I'm sorry I've been a flake this week," she texted on March 22. "I'm flying out to Puerto Rico tomorrow evening."

A number of her friends who shared an interest in tech startup and cryptocurrency and investing were moving to Puerto Rico. She was also tired of the Bay Area's chilly summer evenings, and had noticed a strained relationship with one of her housemates. (Haugen later learned that the woman had picked up on her Facebook information collection and had begun telling mutual acquaintances that she believed Haugen was an undercover federal agent.) Since Facebook was allowing its employees to work from anywhere during the pandemic, giving Puerto Rico a try seemed relatively low stakes. She could always change her mind and come back to California.

I didn't hear from Haugen again for a couple of weeks, until she called with some bad news. After she had flown to Puerto Rico and rented an apartment in San Juan, she had updated her address in Facebook's payroll system, prompting a call from Human Resources. While the company could support employees relocating to other countries during the pandemic, the vagaries of payroll taxes in U.S. territories made her continued employment in Puerto Rico impossible. She would need to promptly return to the States or resign by mid-May at the latest, a little over a month off.

Already questioning her stamina as an internal Facebook mole, she decided to stay in San Juan. Whatever information gathering she was going to do would have to be finished before her final day at Facebook.

"I would like for us to figure out a way to do things remotely," Haugen texted me.

We tried that for a week. But Haugen was clearly exhausted. As much as she was a one-woman operation, having company was helpful. So, with the encouragement of my editor, Brad Reagan, I floated the idea of joining her in San Juan, just for a few days. She accepted.

"So are you here for crypto, too?" asked the guy with whom I was splitting a cab from Luis Muñoz Marín Airport into the city. He had sold his house and business in Pasadena in late 2020 and, in defiance of every principle of financial management, dumped the full proceeds into Bitcoin, Ethereum, and a couple more obscure cryptocurrencies. His portfolio doubled in just two months.

He was now letting the bet ride, staying at a waterfront hotel and waiting to see what happened next. If crypto kept going up, he was going to buy a mansion. If it went down, he said, he'd end up living "in a little box" somewhere inland. After I gave a noncommittal answer as to why I was in town, he told me that, if I was here to write about crypto, I should come meet him for breakfast.

I never saw the guy again after he got out of our cab in San Juan's Condado neighborhood, but the brief encounter left an impression. Among the mainland's newly arrived expats, there was an impressive tolerance for risk.

I found Haugen unfazed by her pending unemployment and change of locale. After taking a short vacation, she was spending most of her time with her laptop at a small table surrounded by unopened packages that she and a new crop of roommates had shipped to themselves before moving to San Juan. Three hours ahead of the West Coast, she planned to spend each morning documenting what she could and then put in a regular workday when her team in Menlo Park came online. In the evenings, we would meet for dinner and talk.

Her routine looked a lot like it had in San Francisco; she just put on a higher-SPF sunblock when she went outside. This intentionally boring lifestyle was a conscious nod to her Midwestern sensibility, Haugen told me. Some of the nascent crypto tycoons she knew were celebrating too hard. Haugen's life didn't need to get any weirder than it already was.

Haugen and I had long ago discussed that, if she actually pulled as much material as she was contemplating, she would almost certainly get caught by Facebook. Even assuming she didn't hit any internal trip wires set up to catch leakers, the eventual publication of stories about the material would end up outing her. Each time Facebook learned that a specific document had been leaked, it could look and see who had accessed that information and when, slowly whittling down a list of possible suspects.

She was, at least in principle, okay with that eventuality. But her goal was to survive as long as she could, ideally leaving the company before anyone wised up. To that end, she had largely documented topics directly relevant to her job and avoided rummaging. Consequently, neither she nor I had a good sense of what would be accessible to her once she started exploring Facebook's broader network of Workplace groups, online documents, tracking metrics, and incident response tickets.

The only reasonable assumption was: not much. After leaking had spiked following employee discontent over the company's response to Trump's "looting and shooting" post, Facebook had begun locking down its systems. Forums that were once viewable by anybody with an employee ID were becoming invitation-only, and newly hired internal moderators would sometimes disappear controversial posts from Workplace.

Workplace was a tough system to navigate, shockingly resistant to keyword searches. Results would turn up a few things, but finding a specific item usually required either an encyclopedic memory or fifteen minutes of puttering around adjacent discussion forums. A particular item of research might be accessible via unlabeled hyperlinks in a half-dozen different posts and documents, with the sheer volume of unstructured, cross-posted material making a clean inventory impossible. Much of this could be attributed to vestiges of Facebook's historic culture of openness, but some of it was pure fuckup. Documents subject to supposed attorney-client privilege and draft presentations to Zuckerberg, complete with the full history of senior executive edits, were occasionally posted in places where north of 60,000 employees, not to mention an unknown number of contractors, could view them.

Facebook's internal communications platform functioned remarkably like its outward-facing product, with a similar layout and features. Groups like "Wrong Answers Only" and "Shitposting@" contained tens of thousands of members, displaying a culture of self-mockery. When the company's chief engineer denounced employees who were "coasting" during remote work, employees proudly shared pictures of company-branded drink coasters.

Like its public-facing sibling, the internal platform sometimes got spicy. People were almost always polite—they were interacting with colleagues, after all—but frustrations were often aired in comments sections rather than diplomatically worded emails.

Employees on the way out the door could be especially scathing. The company had a tradition of "badge posting," in which departing staffers would combine a picture of the employee ID they were about to turn in with their parting thoughts to colleagues. Amid expressions of gratitude and invitations to keep in touch, employees sometimes explained why they had chosen to quit.

The reasons could be revelatory. In 2016, the *New York Times* had reported that Facebook was quietly working on a censorship tool in an effort to gain entry to the Chinese market. While the story was a monster, it didn't come as a surprise to many people inside the company. Four months earlier, an engineer had discovered that another team had modified a spam-fighting tool in a way that would allow an outside party control over content moderation in specific geographic regions. In response, he had resigned, leaving behind a badge post correctly surmising that the code was meant to loop in Chinese censors.

With a literary mic drop, the post closed out with a quote on ethics from Charlotte Brontë's *Jane Eyre:* "Laws and principles are not for the times when there is no temptation: they are for such moments as this, when body and soul rise in mutiny against their rigour; stringent are they; inviolate they shall be. If at my individual convenience I might break them, what would be their worth?"

Garnering 1,100 reactions, 132 comments, and 57 shares, the post took the program from top secret to open secret. Its author had just pioneered a new template: the hard-hitting Facebook farewell.

That particular farewell came during a time when Facebook's employee satisfaction surveys were generally positive, before the time of endless crisis, when societal concerns became top of mind. In the intervening years, Facebook had hired a massive base of Integrity employees to work on those issues, and seriously pissed off a nontrivial portion of them.

Consequently, some badge posts began to take on a more mutinous tone. Staffers who had done groundbreaking work on radicalization, human trafficking, and misinformation would summarize both their accomplishments and where they believed the company had come up short on technical and moral grounds. Some broadsides against the company ended on a hopeful note, including detailed, jargon-light instructions for how, in the future, their successors could resurrect the work.

These posts were gold mines for Haugen, connecting product proposals, experimental results, and ideas in ways that would have been impossible for an outsider to re-create. She photographed not just the posts themselves but the material they linked to, following the threads to other topics and documents. A half dozen were truly incredible, unauthorized chronicles of Facebook's dawning understanding of the way its design determined what its users consumed and shared. The authors of these documents hadn't been trying to push Facebook toward social engineering—they had been warning that the company had already wandered into doing so and was now neck deep.

Haugen spent most mornings gathering documents and the rest of the workday finishing up her actual tasks at Facebook, though she was beginning to sneak research sessions in during Zoom calls. Long afterward, her colleagues would tell me that her investigations included asking what were, in hindsight, unusual questions, such as, Did they recall where to find the link to their work documenting the prevalence of hate speech in Amharic, the most widely spoken language in Ethiopia?

I spent most days trying to understand what she had gathered, and most evenings we would meet for dinner and drinks and talk over what it all added up to. Haugen was averaging a few hundred screenshots a day. From what I could tell, they were a blurry mess, but almost always legible.

Working with a source siphoning documents is a tricky business, with ethical and legal considerations intertwined. I could not ask her to gather specific documents, nor could I even lay a finger on her keyboard, a move that the *Journal*'s lawyers made clear could potentially open me up to charges under the Computer Fraud and

Abuse Act, the federal law under which most hacking crimes get prosecuted.

That left me with a limited mandate. My work consisted of discussing whatever information Haugen wanted to share, researching whatever was known about the topic outside of the company, and then asking follow-up questions. The broader picture that emerged was not that vile things were happening on Facebook—it was that Facebook knew. It knew the extent of the problems on its platform, it knew (and usually ignored) the ways it might address them, and, above all, it knew how the dynamics of its social network differed from those of either the open internet or offline life.

Between Haugen's own sense of Facebook's problems and months of talking, we had already agreed on some basic questions we wanted to answer. What did the company think about its effects on politics? What were its platforms' capacities to detect problem content? How did Facebook interact with political figures and high-profile entities?

After four or five days in San Juan, I had learned enough to compellingly argue to my editor that my trip wasn't just deep cover for a beach vacation. I went to Marshalls to buy clean clothes and rented a vacation condo a short walk from Haugen's apartment. We would use the apartment as our joint office, eager to avoid exposing her roommates to a one-woman espionage operation. I would stay until Haugen quit, got fired, or wanted to stop.

Gathering the documents took time. That was partly because Haugen had a whole new language to learn. How did enforcement and investigative tools named CORGI, Bonjovi, Rabbithole, Drebbel, Black Hole, and Bouncer work? She knew some, but not all, of the systems and jargon.

Some searches were the result of a whim. Haugen's actual work at Facebook focused on networked misinformation. But we had talked over questions about Instagram's effects on teenage mental health. Could there be anything there?

Haugen spent a half hour searching and didn't find much. That evening, she tried again and turned up a few documents. Among them was a 2019 presentation by user experience researchers finding that, while causality was hard to establish, Instagram's aesthetic of casual perfection could trigger negative thinking among some users. The researchers' best understanding was summarized this way: "We make body image issues worse for one in three teen girls."

Holy shit.

I was astonished not just by the finding but by the fact that we'd found it at all. Facebook had been tightening up controls for access to sensitive material for more than a year—yet stuff like this was still sitting there, viewable by anyone who poked around. Somehow, before Haugen, no one had.

Splitting a bottle of wine on the balcony of my condo one evening, we asked each other how it was possible that she hadn't been stopped. At minimum, this was a monumental failure on the part of Facebook's Security team. A core tenet of cybersecurity is to be on alert for abnormal usage, and what Haugen was doing was far from ordinary. On my second weekend in Puerto Rico, she had thrown caution to the wind and collected documents on both Saturday and Sunday. Anyone monitoring the volume of this soon-to-resign employee's activity might have raised an eyebrow at how much information she was consuming. If they had noticed that an employee assigned to hunting state-sponsored manipulation had begun ferreting out documents on teen mental health, they would have hit the panic button.

This was, ironically, the same lack of concern that Facebook had exhibited toward users who overused its platforms. The company had failed to notice or care that a tiny number of people were making friends, sending out invitations, and resharing content in ways that correlated with bad outcomes. *More usage was good*—why would Facebook apply a different standard to Facebook Workplace than to its public platforms?

Facebook's leak defenses were predicated on the bet that people like Haugen, with well-paid desk jobs, wouldn't throw away a golden ticket, or risk the possibility of a lawsuit from a trillion-dollar company. She was contemptuous of that thinking from the start. Compared with the dangers she believed Facebook posed to users overseas in particular, the risk of screwing up a cushy career didn't rank.

Still, every morning Haugen woke up scared that she would find that a meeting with Facebook's Security team had been booked on her calendar. She asked whether I thought Facebook would sue her. I could truthfully say that there was no record of Facebook going to court over whistleblower violations. I had to include a caveat: no Facebook employee had previously undertaken a six-month espionage campaign while working on a team devoted to counterespionage. Haugen said she was ready to accept whatever consequences might come, but the frequency with which she brought up the risks was a stark reminder of which one of us had more on the line. She began working to get her expired passport renewed because, well, one never knows.

As one week turned to two and then three, with no sign that the company would step in to stop her before her final day in mid-May, a new topic of conversation between us emerged. If she failed to grab something important in her remaining time at the company, it was likely that nobody else would ever have the chance. She had already taken 10,000 screenshots, amassing a horde of information that surely amounted to the largest leak in the company's history. It was bizarre that corporate security was as loose as it was, and flatly implausible that it would stay that way once the information she had gathered had been revealed.

I liked this topic of conversation. At a time when Haugen had taken to wearing a wrist brace due to the repetitive strain of scrolling and photographing her screen, it was a reminder that the window was closing. If I explicitly said as much, however gently, it would feel exploitative. Haugen was clearly exhausted and beginning to complain that the medication she used to control her neuropathy pain wasn't cutting it.

There was also an argument to make that ensuring the scale of her transgression was as large as possible would bring a form of protection. For Facebook to go after a leaker would be one thing; for it to pursue a potential star whistleblower, another.

I had been trying to track what information Haugen was learning, a task that became only more daunting as she filled up the memory card of her cheap phone twice a day. While she was being judicious in what she recorded, both of us were getting hazy on the exact contours of what she had searched for and grabbed. Since a full inventory would have to wait until after she was gone, it was better to be safe than sorry.

To make best use of our time during Haugen's final week at the company, we converted my vacation studio into a half-assed office for two. Haugen took the table and I took the bed. If she needed privacy for a call with colleagues, I would go out onto the balcony, though I sometimes stuck around for less sensitive conversations, including some related to her departure. When she had interviewed with Facebook, Haugen had told recruiters that she wanted the job so she could help the company fix its products. On the way out, she told managers that she was leaving because the company wouldn't let her.

"I have met very conscientious, sincere, earnest people who are trying to solve very important problems and I feel honored that I got to work with them," she told a manager during one such video call. "But the fire drill culture where things don't get fixed until the press reports on them is deeply toxic."

The days grew longer as our time grew short. I would fetch coffee before Haugen arrived in the morning, then order food delivery for our meals. We took our lunch breaks on the balcony, and when we could see manatees grazing on seagrass in the Condado Lagoon below, we would joke that they were her efforts' mascot. Haugen brought over a portable speaker so she could play dance music at night when we got tired. Of her final forty-eight hours at Facebook, we worked thirty-eight.

"Good work, consistently, over a long period of time," Haugen said.

Haugen was due to lose access to Facebook's network at 7:00 p.m. on May 17 and I had made a reservation for that time at an upscale restaurant to celebrate. I called a cab for 6:30 p.m. to take us there, but Haugen was still busy downloading the company's entire organizational chart, an especially delicate task that she had chosen for last. I went downstairs, paid the driver $20 to not leave, and then went back up to my apartment to hurry her out the door.

Before she closed her laptop, she entered one final search query into Workplace, assuming it would be the final thing that Facebook's Security team would see in the inevitable forensic review.

"I don't hate Facebook," it began. "I love Facebook. I want to save it."

She finished typing, pressed enter, and closed her laptop.

A day later, I flew back to the West Coast and got to work.

17

Facebook had battled a series of scandals—over its approach to data, to politics, to extremism percolating on its platforms—but there had always been a low-level hum of discontent from people who said using it didn't make them feel good.

Internal research done in 2019 found that a little over 3 percent of American users were suffering from "serious problems with sleep, work, or relationships that they attribute to Facebook" and felt anxiety about their relationship with the product. The research suggested that roughly 10 million Americans suffered from "problematic use" of the main Facebook platform alone. "Though Facebook use may not meet clinical standards for addiction, we want to fix the underlying design issues that lead to this concern," the researchers wrote.

With the country still in the grips of the COVID-19 pandemic, mental health—and particularly the mental health of teens, spending their formative years learning from home instead of in school with their peers—had begun to receive even more attention from officials, the media, and parents. That the self-isolation and quarantining prescribed by the pandemic pushed ever more people online more often only compounded the issue.

Facebook never flatly denied that its products might be bad for teenagers, but it regularly sparred with those who said they were. There was enough public disagreement and doubt about the mental health effects of social media that executives could point to good-faith skepticism. Survey studies were unreliable, Instagram's Mosseri would note in public forums, saying that a widely cited Oxford study found that users who said social media was a problem for them often overestimated their own use. Mental health was subjective and causality difficult to prove.

When Zuckerberg was called in to testify before Congress in March 2021, alongside then-CEO of Twitter Jack Dorsey and Google CEO Sundar Pichai, about the tech companies' role in the January 6 riot, Representative Cathy McMorris Rodgers asked him whether social media might be contributing to rising rates of depression among teenagers, and about social media's effect on mental health overall. He declined to meaningfully engage. "I don't think that the research is conclusive on that," he replied.

The research might not have been conclusive, but it was extensive. In 2020, Instagram's Well-Being team had run a study of massive scope, surveying 100,000 users in nine countries about negative social comparison on Instagram. The researchers then paired the answers with individualized data on how each user who took the survey had behaved on Instagram, including how and what they posted. They found that, for a sizable minority of users, especially those in Western countries, Instagram was a rough place. Ten percent reported that they "often or always" felt worse about themselves after using the platform, and a quarter believed Instagram made negative comparison worse.

Their findings were incredibly granular. They found that fashion and beauty content produced negative feelings in ways that adjacent content like fitness did not. They found that "people feel worse when they see more celebrities in feed," and that Kylie Jenner seemed to be unusually triggering, while Dwayne "The Rock" Johnson was no trouble at all. They found that people judged themselves far more harshly against friends than celebrities. A movie star's post needed 10,000 likes before it caused social comparison, whereas, for a peer, the number was ten.

In order to confront these findings, the Well-Being team suggested that the company cut back on recommending celebrities for people to follow, or reweight Instagram's feed to include less celebrity and fashion content, or de-emphasize comments about people's appearance. As a fellow employee noted in response to summaries of these proposals on Workplace, the Well-Being team was suggesting that Instagram become less like Instagram.

"Isn't that what IG is mostly about?" the man wrote. "Getting a peek at the (very photogenic) life of the top 0.1%? Isn't that the reason why teens are on the platform?"

Meanwhile, the company was funding an array of mental health nonprofits that spread the message that Facebook was engaging with the issue, and it talked up the ways in which users could empower themselves on the company's platforms—as though the problems were not baked in. One such organization, the National Eating Disorders Association (NEDA), promised to help users "ensure your time on Instagram is healthy, supportive, and empowering," acknowledging that harmful content existed on social media, but assuring the public that "Instagram has taken direct action to make their platform safer for all users." On its blog, NEDA highlighted testimonials of Instagram users who had recovered from eating disorders thanks to connections made on the platform.

The template was repeated with other groups. An executive at the Jed Foundation, a splashy teen mental health nonprofit that received Instagram funding, told the *Washington Post* that the platform was "dedicated to mental health" and "committed to ensuring that the user is protected." Jed, too, collaborated with Instagram on body-positive marketing campaigns such as "Pressure to Be Perfect," which portrayed being "mindful" as an antidote. As part of its company-funded work, the group encouraged teens to undertake "daily affirmations" of the phrase "I am in control of my experience on Instagram."

Rather than acknowledging or confronting defects in the platform's design, the problem was presented as one of awareness and self-control. The concept didn't seem unreasonable on its face, especially with teen mental health nonprofits as the messenger. And it happened to be a perfect fit for the company's final decision on Project Daisy, its proposed plan to hide likes on Instagram in order to improve users' experience.

Things had stalled after Mosseri and other executives had presented the plan to Zuckerberg in early 2020. The company had done research that found that, while users did like the idea of hiding likes, it didn't lead them to share more and the move didn't lead to a change in "overall well-being measures," as a presentation to Zuckerberg by Mosseri, Alex Schultz, and other executives at the time put it. There were other drawbacks, too. Without a like count to signal which posts were popular, users were spending a little less time on the app and clicking on fewer ads. Revenue might fall as much as 1 percent.

As an effort to "depressurize sharing and to reduce social comparison," Project Daisy had been a bust. But, back then, the presentation said the company might want to go through with hiding likes anyway. "A Daisy launch would be received by press and parents as a strong positive indication that Instagram cares about its users, especially when taken alongside other press-positive launches," it said.

Hegeman, the head of News Feed, pushed back, questioning in notes on the draft presentation whether viewing popular posts on Instagram could truly be said to cause users to feel bad about themselves. "I realize this is a plausible hypothesis," he wrote, before adding that definitely stating causality "feels like a bit of a jump." A top researcher pushed back, citing qualitative work in which users had clearly expressed that they felt bad about themselves when they saw popular posts. Hegeman removed the line from the presentation anyway. With the approval of the company's head of Research, Pratiti Raychoudhury, the statement of a causal contributing role was not included in the presentation to Zuckerberg.

The evidentiary standard being applied was remarkable in comparison to the company's usual requirements. Facebook executives from Zuckerberg down regularly spoke about the platform's upside for well-being. It certainly hadn't demanded causal research proving that its product was good for people. But suggesting the downside demanded proof positive.

The project was shelved until March 2021—when *BuzzFeed News* obtained a leaked memo revealing that Instagram had plans to build a platform for preteens.

The report set off a furor. Children under the age of thirteen were banned from the platform. Just two days prior, Instagram had published a blog post outlining steps it was taking to make Instagram safer for teens, including a new guide for parents, restricting the ability of adults to send DMs to kids under eighteen, and encouraging teens to make their accounts private.

The outcry was immediate, loud, and long-lasting. Within two months, attorneys general from forty-four states and U.S. territories would write a letter to Facebook urging it to abandon the plan, citing both privacy concerns and worries over mental health effects.

Given the backlash, perhaps it was time to revive Project Daisy after all. Facebook took the product into additional testing and settled on a watered-down version that did not hide likes entirely, but offered users the ability to opt in to not seeing them.

The move was modest at best. Hiding likes hadn't been hugely effective in the first place, and only a small fraction of users used opt-in features anyway. But, in keeping with the strategy laid out in the presentation to Zuckerberg the year before, Mosseri went on *Good Morning America* to announce the change.

After Mosseri made his "big announcement," as host Gayle King called it, he explained that the change would allow users to "focus more on the people they care about and being inspired." He then walked viewers through the process of switching the feature on if they wished.

"You've given us the choice and we appreciate that," King concluded.

Mosseri made other efforts to promote the change, including a conference call with reporters in May. Back home in Oakland, I dialed in and, when the time came for questions, told Mosseri I would be putting aside Project Daisy to instead ask: What did Instagram know about its platform's ability to affect its users' sense of well-being?

"That's a good question," Mosseri responded. Instagram's effects were likely similar to those of other social media platforms, he said, "small effects positive, and small effects negative—but quite small."

There was nothing in the public record to contradict that self-appraisal, nothing to anyone else listening that would suggest the platform was anything less than safe in its current form. But I had gotten back from Puerto Rico the week before, and I was sitting on years of internal findings to the contrary. Haugen had shared with me a risk investigation conducted just two months prior that found that Instagram was reliably recommending pro-anorexia content to users who seemed vulnerable to it. The finding was in keeping with everything the company knew about the tendency of its recommendation systems to steer users toward edgy content that it was incapable of policing.

"We are practically not doing anything," the researchers had written, noting that Instagram wasn't currently able to stop itself from promoting underweight influencers and aggressive dieting. A test account that signaled an interest in eating disorder content filled up with pictures of thigh gaps and emaciated limbs.

The problem would be relatively easy for outsiders to document. Instagram was, the research warned, "getting away with it because no one has decided to dial into it."

Arturo Bejar had seen things from a different vantage point, but his conclusions were the same. After returning to Facebook in 2019 following a four-year break, Bejar was working as a consultant on Instagram's Well-Being team—a move inspired by his teen daughter's stories of abuse on the platform. Once filled with optimism about the possibilities of tech, and about Facebook in particular, Bejar's view had become decidedly darker.

As Facebook's original "Mr. Nice," Bejar had pioneered the company's approach to improving the experience of its users. His Compassion team had done particularly groundbreaking work on suicide, creating a way for users to report when they were worried a friend on Facebook was at risk of self-harm, directing them first to regionally appropriate crisis resources and, if there was cause for imminent concern, prompting them to submit a report. A person who appeared depressed would receive a notice the next time they logged on saying, "Someone is worried about you." If someone seemed at immediate risk, Facebook's moderators would escalate it and potentially report it to law enforcement.

"There were a couple of isolated cases where somebody posted on Facebook that they were gonna kill themselves, and then they did it," recalled a member of the team working on suicide prevention. These deaths prompted both introspection and analysis. Sometimes, reviews would reveal that company moderators had dropped the ball. Other times, users had failed to report the risk. And even when users did flag someone in crisis, response times were often slow because the review queues were clogged with bad-faith reports.

Press coverage of a wave of livestreamed suicides in 2017 didn't differentiate between these scenarios. If someone killed themselves on Facebook, that was Facebook's fault. The company had turned to AI to help. To build a classifier for self-harm or graphic violence, the team fed its machine learning tool, FB Learner, a sizable data set, initially text only. The new system showed promise, prioritizing reports so that moderators reached the valid ones roughly twenty times faster. They then turned to live video. What they produced wasn't perfect, but it was good.

While the number of lives the tool saved was unknowable, the team could track how many times the classifier, rather than users, detected suicidal behavior worrying enough that Facebook's Customer Care team reported it to law enforcement. "It was a measure of how many times we were the only hope and we tried, and it was a much bigger number than we expected," said the team member, who began to cry as he recalled the interventions. "We don't know how many of them died, but we tried to save them."

This was Facebook at its best, cobbling together technology in imperfect but powerful ways. The effort had remained in force after Bejar left the company in 2015, but other projects he pioneered weren't as long-lasting. There was the user report submission process that he had fine-tuned, something he believed essential to creating and maintaining a good platform—that was gone. The company had shifted from handling bad user reports to actively discouraging them.

At the same time, Facebook increased its focus on reducing the existence of narrowly defined "bad content" after asking users to

report only clear-cut violations of specific Facebook rules, something it could lean on machine learning to do. Spotting users who
were selling live animals, distributing child sexual abuse material,
or recruiting participants for a terrorist attack was fairly straightforward. Modestly improving a classifier's capacity to detect a woman's
nipple, for example, could prevent tens of millions of "regrettable"
views.

What such automated enforcement could not do was engage
with problems that had subjective components, from hate speech to
bullying. Through a combination of its overreliance on automation
and underreliance on human review, the company did not have the
deftness to referee which posts about anorexia subtly encouraged it,
when referencing someone's religion was inappropriate, or whether
the comment "I love your makeup!" was a compliment or a form of
harassment.

There wasn't much users could do to register discontent with
what they saw, beyond submitting reports that often went nowhere,
or else blocking a user. For years, users had asked for something
like a "dislike" button, but the idea had never appealed to leadership, including Zuckerberg. "Our product instincts told us that a
'dislike' button was ripe for increasing negativity," Julie Zhuo, vice
president of App Design, wrote in a 2021 essay on good product
design.

Driving much of this was the reluctance of Facebook, a company with an otherwise insatiable appetite for data, to collect negative feedback of the kind Bejar had once sought to elicit. Multiple
attempts to push back failed. "Negative sentiment doesn't belong on
FB because we're a happy place where you connect with friends,"
one former director told me, summarizing the company's thinking.
"If we allow you as a user to receive negative feedback, that's not
good for you."

During any given week, Facebook collected 80 billion different
positive signals about what users did like on the platform, but just
0.5 percent as many signals about what they didn't, one data scientist found in a 2019 report. The data scientist noted that the official numbers regarding content didn't offer a clean picture of what

users experienced. Officially prohibited nudity accounted for only 0.05 percent of posts viewed on Facebook, while hate speech and violence stood at around 0.2 percent. But if you lumped in the "borderline" cases, the proportion of problem content ballooned to around 10 percent of all posts viewed. And if you also considered engagement bait, scraped content, and "shocking health information" to cause bad experiences, that number rose to more than 20 percent of all posts viewed on Facebook.

Years later, the company would begin experimenting with introducing simple ways for users to express displeasure, functionally acknowledging that the researcher had a point. But, at the time, Facebook wasn't interested in gathering such information. Because the company designed its platform to collect only positive feedback, it heard only from people who actively enjoyed it.

Bejar shared the data scientist's bafflement that the company wasn't investing heavily in collecting a broader range of data. He spent a great deal of time explaining that not all the uses of negative signal even had to be negative. During his first stint at Facebook, the Compassion team had shown that apparent social boors often improved their behavior when privately informed they had been irritating others. When Facebook told a user that friend-spamming attractive strangers wasn't going over well, there was a fifty-fifty chance they would cut it out without any discipline or account restrictions. What wasn't to like?

The fact was that, outside of clear abuse, Facebook didn't like to consider even the possibility of unhealthy activity. Bejar might consider successfully prodding users to stop sending out random friend requests to be an unequivocal victory, but the Friending team sure as hell didn't. Who was Facebook to say that users should connect only with people they knew?

Bejar understood that this was a design choice. Users could hit the gas on content, but only the platform could tap the brakes. Zuckerberg and other executives had long extolled the virtues of counterspeech as a way to "elevate the dialogue beyond the reach of fear, hate, and violence," as one company announcement loftily put it. As nice as the thought was, the company couldn't claim in good faith that it worked. When a user argued with a page posting a vaguely racist cartoon, the only thing "elevated" was engagement.

Once Bejar recognized what the company was doing, he saw it everywhere. Rather than asking users if they would prefer to see fewer photos of influencers enhanced with filters, Instagram would prefer they express their body-positive feelings by adding to the platform's hundreds of millions of #nofilter posts.

"A reason why people put this beauty and wellness content with exaggerated Photoshop everything is that it gets the positive signal they're looking for," Bejar said. "There was no negative signal that privately allows people to say 'That's kind of gross' so that the platform can downrank it."

It had taken Bejar a year to figure this all out, but he had found his answer to what had gone wrong on Facebook's platforms. Despite its stated goal of respecting its users' wishes, the company was governing Facebook and Instagram according to its own preferences, not theirs.

The realization led to something like an existential crisis for Bejar. "You're told you're a wizard, that you'll find the right answer, that the rest of the world just doesn't get it," he told me. "I'd bought into that ever since I started in Silicon Valley, and when I looked back, I felt shame."

Bejar was not ready to give up on the company, though. He decided he would devote himself to figuring out how to bring internal attention to the company's user experience blind spot. One option was improving TRIPS (Tracking Reach of Integrity Problems Survey), a survey that regularly asked users about their perceived exposure to various types of problem content. Previous efforts to make TRIPS a priority had failed. "That road was littered with a lot of bodies," Bejar said.

Bejar suspected a reason for this failure. Facebook's enforcement efforts were largely irrelevant to users. Efforts to combat the tiny percentage of officially rule-breaking content didn't matter when most of what bothered people wasn't in that category. A report of bullying was lived experience, not an allegation that Facebook could falsify. When someone felt that the content on Instagram had made them feel insecure about their body or their life, the correct response wasn't to question the causality.

From his earlier years of working with the company's senior leadership, Bejar knew better than to push them to embrace the subjectivity of human experience. To have a shot at getting Zuckerberg and other top executives to address the company's blind spot, he would need solid data that quantified the problem.

This was the origin of BEEF, short for "Bad Emotional Experience Feedback." In consultation with a group of like-minded engineers and product managers, Bejar built out a weekly survey intended to quantify the unpleasantness of Instagram, whether it was unwanted exposure to violence or seeing a peer bullied. If the company knew what portion of teenagers, like his own, had to fend off aggressive sexual solicitation and verbal abuse on Instagram every week, perhaps executives would realize how far off the mark its approach to governing the platform was.

"I was holding on to the idea that Mark and all these people just didn't know," Bejar said.

In mid-2021, Bejar was nearing the end of his two-year contract and preparing his final push at persuading Facebook's executives. Bejar's history and the backing of Instagram Well-Being team leaders earned him a shot at presenting his work directly to the company's top executives.

He began the presentation by noting that 51 percent of Instagram users reported having a "bad or harmful" experience on the platform in the previous seven days. But only 1 percent of those users reported the objectionable content to the company, and Instagram took action in 2 percent of those cases. The math meant that the platform remediated only 0.02 percent of what upset users—just one bad experience out of every 5,000.

"The numbers are probably similar on Facebook," he noted, calling the statistics evidence of the company's failure to understand the experiences of users such as his own daughter. Now sixteen, she had recently been told to "get back to the kitchen" after she posted about cars, Bejar said, and she continued receiving the unsolicited dick pics she had been getting since the age of fourteen. "I asked her why boys keep doing that? She said if the only thing that happens is they get blocked, why wouldn't they?"

Two years of research had confirmed that Joanna Bejar's logic was sound. On a weekly basis, 24 percent of all Instagram users between the ages of thirteen and fifteen received unsolicited advances, Bejar informed the executives. Most of that abuse didn't violate the company's policies, and Instagram rarely caught the portion that did.

"Policy enforcement is analogous to the police, it is necessary to prevent crime, but it is not what makes a space feel safe," Bejar wrote. As much as Facebook wanted to fix integrity problems through reducing the prevalence of officially bad content, the company would need to reorient toward experience-based metrics like BEEF and efforts to instill social norms.

"I am appealing to you because I believe that working this way will require a culture shift. I know that everyone in M-team deeply cares about the people we serve, and the communities we are trying to nurture," Bejar wrote.

Bejar had also tried to present the message to executives individually, explaining that there was little correlation between the behavior that Facebook treated as a problem and what its users experienced as one. Mosseri had struck him as supportive and Sandberg as empathetic—if unengaged—on the subject of abusive behavior toward young women. The surprise had been Cox.

"He knew," Bejar said. Facebook's chief product officer appeared to grasp Bejar's argument: the company's platforms didn't let users meaningfully object to how they were treated, much less give them the tools to establish the expectation of something better. But after hearing Bejar's proposals, including that perhaps the company should begin urgently working to reduce the proportion of teenagers who reported unwanted sexual content each week, he was noncommittal. "Oh, yeah, that sounds really interesting," Bejar recalled Cox telling him. "Let me kick it over to Guy Rosen."

Rosen, the architect of Facebook's metrics-driven enforcement work, clearly lacked both the interest and the clout needed to push a proposal like this. The brush-off floored Bejar. "Cox was one of the people that had my back when I was at the company. He really got the compassion stuff, and he championed it to Mark." (A company spokeswoman said that Cox's referral simply reflected his belief that Rosen, as head of Integrity, was the right executive to discuss the matter.)

Perhaps it had been a long shot to think that anyone could have persuaded Zuckerberg to rethink how he had built Facebook. Without Cox's support, it was a lost cause. The company would eventually roll out a few minor features along the lines of what Bejar had been thinking, but he considered them to be tepid and PR-focused. Restricting adults' ability to initiate private conversations with kids after they had been repeatedly blocked by underage users was vaguely remedial, not a leap forward.

By October, Bejar's contract was up. Even if it hadn't been, there wouldn't have been a point to staying on.

"The machine would just keep on working the way that it was working," he said.

Around the same time as Bejar's second departure, Brandon Silverman was also hitting the road, though for a different reason. By allowing reporters, advocacy groups, and academics to track and study viral content, CrowdTangle had proven itself a bit too useful, allowing outsiders to detect the same quality and algorithm-gaming problems that Silverman had internally flagged for senior executives.

Exposure sometimes altered major decisions. When a COVID conspiracy film titled "Plandemic" exploded on Facebook and rival platforms in May 2020, Kaplan and the Public Policy team blocked the Health team from immediately removing it, until a *New York Times* reporter tweeted that CrowdTangle showed "Plandemic" was Facebook's number one post. The Policy team reversed itself and approved the takedown with no further discussion.

Even the threat that someone might be watching could focus leadership attention. Whenever Silverman went to an executive with a dashboard showing that Facebook was pumping out especially embarrassing or potentially harmful content that day, the first question was invariably, *So anybody can see this?* When informed that they could, the PR fire-in-waiting became a priority.

If transparency had its costs and benefits, nothing highlighted the costs better than a Twitter bot set up by *New York Times* reporter Kevin Roose. Using methodology created with the help of a Crowd-Tangle staffer, Roose found a clever way to put together a daily top ten of the platform's highest-engagement content in the United States, producing a leaderboard that demonstrated how thoroughly partisan publishers and viral content aggregators dominated the engagement signals that Facebook valued most.

The degree to which that single automated Twitter account got under the skin of Facebook's leadership would be difficult to over-state. Alex Schultz, the VP who oversaw Facebook's Growth team, was especially incensed—partly because he considered raw engage-ment counts to be misleading, but more because it was Facebook's own tool reminding the world every morning at 9:00 a.m. Pacific that the platform's content was trash.

"The reaction was to prove the data wrong," recalled Brian Boland. But efforts to employ other methodologies only produced top ten lists that were nearly as unflattering. Schultz began lobby-ing to kill off CrowdTangle altogether, replacing it with periodic top content reports of its own design. That would still be more transpar-ency than any of Facebook's rivals offered, Schultz noted.

Even before pissing off one of the company's top executives, CrowdTangle had been vulnerable. Because Silverman's team worked with so many different product groups, it lacked a single executive advocate or a steady budget. Boland had made a case for the value of CrowdTangle's work, sending a note at the end of 2020 to his roughly four hundred fellow Facebook VPs urging them to join him in supporting it and other efforts to share data with outside researchers. Just as Facebook provided advertisers with data about the effectiveness of their marketing campaigns, Boland argued, the company ought to give the public enough information to decide whether criticisms of its platforms were valid. "You can choose to invest in understanding these areas with your 2021 headcount allocation," Boland wrote. "Don't wait for someone to do this centrally if you find it important—we aren't on a path for that."

With Facebook bruised from the 2020 election and the Capitol riot, however, Schultz handily won the fight. In April 2021, Silverman convened his staff on a conference call and told them that CrowdTangle's team was being disbanded. The tool itself would live on, at least for a while, but under the control of Rosen's Integrity team. Efforts to build new features ended.

"Brandon caught fire on building transparency for Facebook," Boland recalled. "He wanted to make Facebook data public in an easily digestible way around the globe, so researchers and journalists could access it."

Boland was already gone by the time the CrowdTangle team got disbanded. When Chris Cox had returned to Facebook in the summer of 2020, Boland had gone to him with a proposal for a new team that would respond to external criticism of the company not by attempting to refute it but by investigating whether it was true.

"Chris Cox was the guy who could take it forward, but he'd been back at the company two or three weeks at that time," Boland said. Again Cox was noncommittal. "It died as a whimper." Though outspoken inside Facebook, Boland also left quietly.

"I was still wrestling with how can I believe what I've come to believe about the negative impact of the platform when every senior executive is telling me I'm wrong?" Boland said. Nobody tried to rebut his specific concerns or argue with the internal research he was citing—they just treated the prospect of Facebook being responsible for significant harm as preposterous.

"Boz would just say, 'You're completely off base,'" Boland said. "Data wins arguments at Facebook, except for this one."

As for Silverman, he stayed on a few more months in an effort to find members of the CrowdTangle staff new roles. He left in the fall of 2021.

18

Near the end of July 2021, John Pinette, Facebook's then-head of Communications, stopped by the *Wall Street Journal's* San Francisco Bureau for an awkward casual visit—he must have known something was up. Sitting with him in a large, empty conference room, my editor, Brad Reagan, and our bureau chief, Jason Dean, discussed Facebook's recent feuding with Apple and broad trends in tech coverage, before telling the avuncular Pinette—a former Catholic priest—that we were indeed working on something substantial but weren't prepared to talk about it yet.

It had been three months since I'd flown home from Puerto Rico. Now the *Journal* had a hard drive containing 22,000 screenshots of 1,200 documents that Haugen had collected over the span of six months. Brad recruited seven other reporters to aid in the effort of combing through all of it and doing the extensive additional research and reporting needed before we could publish.

While pulling clean drafts together took thousands of hours of work, the stories had all but revealed themselves. Facebook had allowed human trafficking to take place in the Persian Gulf on its platform as long as it occurred through brick-and-mortar businesses. In trying to improve the platform and boost user numbers, it had actually made the site, and the people who used it, angrier. Mental health researchers had concluded "we make body issues worse" and that Instagram was a toxic place for many teen girls, in particular.

We divided up the stories among ourselves. Georgia Wells began interviewing young women who had developed eating disorders or body image issues of the sort that Instagram's researchers worried their product might aggravate. The story she led would cite company documents that found "comparisons on Instagram can change how young women view and describe themselves," citing research that found 32 percent of teen girls said that "when they felt bad about their bodies, Instagram made them feel worse."

For a story on Facebook's failings in developing countries, Newley Purnell and Justin Scheck found a woman who had been trafficked from Kenya to Saudi Arabia, and they were looking into the role Facebook had played in recruiting hit men for Mexican drug lords. That story would reveal that Facebook had failed to effectively shut down the presence of the Jalisco New Generation Cartel on Facebook and Instagram, allowing it to repeatedly post photos of extreme gore, including severed hands and beheadings.

Looking into how the platform encouraged anger, Keach Hagey relied on documents showing that political parties in Poland had complained to Facebook that the changes it had made around engagement made them embrace more negative positions. The documents didn't name the parties; she was trying to figure out which ones.

Deepa Seetharaman was working to understand how Facebook's vaunted AI managed to take down such a tiny percentage—a low single-digit percent, according to the documents Haugen had given me—of hate speech on the platform, including constant failures to identify first-person shooting videos and racist rants.

And Sam Schechner and Emily Glazer were studying how activists had spread baseless doubts about the COVID vaccine so effectively that Facebook had to reimpose its Break the Glass measures in May 2021—the third time it had done so in the United States in six months.

I chipped in on all these stories, but I spent the bulk of my time focusing on two: revealing the existence of XCheck, Facebook's program to give preferential treatment to VIP users, and then examining its response to January 6.

In Puerto Rico, Haugen and I had discussed the merits of publishing the stories slowly, releasing one damning article each week over the span of months, giving the complex issues in each story the attention they deserved. Senior editors at the *Journal,* unsurprisingly, had other ideas. They wanted stories published daily, dominating a solid week of tech news, a way to clearly demonstrate that the project was something extraordinary.

Haugen was viscerally hostile to the idea, but she didn't get to have a say. I did, however, and I shared her view. I dismissively branded the proposed schedule "Shark Week," in honor of the Discovery Channel's famously craven ratings grab. Turns out I didn't have a say either.

The XCheck story, published with the headline "Facebook Says Its Rules Apply to All. Company Documents Reveal a Secret Elite That's Exempt," ran on September 13, launching a series that would captivate Congress and average users, revealing the inner workings and deliberations of a company that had worked hard to avoid detailed scrutiny. The choice to lead with the XCheck story was intentional. Unlike some of the other stories, this one didn't require explaining anything technical—content recommendation systems, classifiers, the mechanics of virality. The gist was simply that Facebook said it treated everyone the same and it did not.

But it had been a scramble. Since October 2020, Facebook had been working under an Oversight Board, a body that was meant to serve as a sort of independent appeals court for the company's moderation decisions. In practice, the board's authority was limited largely to issuing Solomonic decisions on individual social media posts.

Facebook's decision to indefinitely suspend Trump in the wake of January 6 had been a moderation call like no other, and the company asked the board to weigh in. On May 4, the board gave its ruling: the decision to suspend Trump was justifiable, it found, but the way Facebook made the call was an unholy mess, following no discernible procedure. The board told Facebook it had six months to come up with a proportionate response, namely setting a time limit on the suspension.

The board also took the opportunity to pepper the company

with questions about how it handled misbehavior by public figures. Among them: a demand that Facebook explain a program known as "cross check."

Haugen and I had been in Puerto Rico at the time, and the mention caught our eye. We were knee deep in scathing documents declaring XCheck to be a train wreck. The documents showed that Facebook knew it had a problem. Repeated internal reviews had found its protection for VIPs to be dangerous, indefensible, and mismanaged. The company had, documents showed, erroneously granted protection to "abusive accounts" and "persistent violators" of its rules.

The documents further revealed how Civic had tried to tackle the program's flaws. In response to a June 2020 call by Chakrabarti for "big ideas" on how to make the platform fairer, one of Civic's engineers had argued that Facebook should publicly announce XCheck's failures, as well as which public figures were receiving its protections.

"This big enough for you, Samidh?" the engineer had written, adding a winking emoji.

Chakrabarti looped the manager who was in charge of fixing XCheck into the discussion. The manager acknowledged the program's atrocious state but said the "business risk" of revealing XCheck's failures would be too great. The best the company could do was try to clean up the mess in private.

As grist for a news story, the XCheck program had pretty much everything a reporter could ask for. Facebook's behavior wasn't just indefensible—it was indefensible by the company's own documented admission. Zuckerberg, Monika Bickert, and other executives had long promised to treat all users equally, but that was just talk. Rather than challenging society's elite, Facebook had kowtowed to them. XCheck was, as the company's own review found, a "betrayal" of its 3 billion users.

There was one potentially complicating factor. Other documents gathered by Haugen showed that, in recent months, Facebook had been making a good-faith effort to rein in the program's worst abuses. It had eliminated "whitelist" enforcement exemptions for

the most serious forms of misconduct and halted the mass enroll-
ment of new accounts into XCheck to "stop the bleeding," as one
manager had written. Fully eliminating special treatment remained
unthinkable—the company still intended to give VIPs "the benefit of the doubt." But, by the time the Oversight Board demanded
details on the program, Facebook was at least in a position to say
that reforms were underway.

In that context, the timing posed a dilemma for the *Journal*. The
board's well-targeted questioning provided Facebook an opportunity to highlight the program's "get well plan." Even if Facebook only
vaguely acknowledged the program's past troubles, that would be
enough to hand the Oversight Board a victory and put a dirty secret
to rest.

I desperately wanted to write about XCheck before then—but
doing so would inevitably provoke a leak investigation inside Facebook and potentially tip the company off to Haugen's identity. We
decided to wait.

Holding off was the right call, but that didn't make waiting for
Facebook's reply to the Oversight Board any less agonizing. When
the company issued its response later in May, I read the document
with a clenched jaw. Facebook had agreed to grant the board's
request for information about XCheck and "any exceptional processes that apply to influential users."

Damn it, I thought. Our story was dead. Then I read the details.

"We want to make clear that we remove content from Facebook,
no matter who posts it," Facebook's response to the Oversight Board
read. "Cross check simply means that we give some content from
certain Pages or Profiles additional review."

There was no mention of whitelisting, of C-suite interventions
to protect famous athletes, of queues of likely violating posts from
VIPs that never got reviewed. Although our documents showed
that at least 7 million of the platform's most prominent users were
shielded by some form of XCheck, Facebook assured the board that
it applied to only "a small number of decisions." The only XCheck-
related request that Facebook didn't address was for data that might
show whether XChecked users had received preferential treatment.

"It is not feasible to track this information," Facebook responded, neglecting to mention that it was exempting some users from enforcement entirely.

Whatever program Facebook was describing, it sure wasn't XCheck.

The staffers responsible for drafting Facebook's response to the Oversight Board—lawyers on the Public Policy team—did not consider themselves to have lied, a person familiar with the matter would later tell me. They just didn't respect the board's effort to push beyond adjudicating moderation decisions into adjudicating how Facebook ran its platforms, and so they had been stingy with information accordingly.

"The center of gravity on the board members' résumés is law, human rights, and journalism," the person told me. "That's a limiting factor on the board's credibility internally."

After reading the response, I called Brad to tell him there was still a story in XCheck. Then I called Haugen to tell her that, given the opportunity to get itself out of a mess, the company had dug itself in deeper.

Haugen had always maintained to me that she didn't consider Facebook malicious, just in way over its head. Here was seeming proof of the opposite. Facebook wouldn't even be forthright about its content moderation problems with an institution that it created to help it with content moderation problems.

"I hope this helps you understand why I did what I did," Haugen responded.

I went ahead with reporting the story. I tried to get an interview with Guy Rosen about the program's cleanup but was unsuccessful. Facebook provided us with a statement, saying XCheck did not provide special treatment to the powerful, and whatever problems the program had were being addressed, though it provided no specifics. The company also said that it had not misled the Oversight Board about XCheck in any fashion.

I heard that, internally, Rosen told colleagues that he didn't see what the big deal was. It was only reasonable, after all, that Facebook would treat its VIP users similarly to how airlines treat their most valuable passengers. But the company nonetheless was preparing for the board to be furious.

Facebook evidently went so far as to try to game out how events would play out once the truth of XCheck was revealed. One of several teams responsible for liaising with the Oversight Board set up a "Murder Board," a committee of employees tasked with role-playing a body of academics, former politicians, and human rights lawyers enraged about having been misled.

Facebook need not have worried. When we approached the Oversight Board with a request for comment, telling them about internal documents showing that XCheck had exempted some VIP users from content moderation and given favorable treatment to others, a spokesperson declined to engage. Only after we published the story did the board gently speak up, issuing a press release titled "To Treat Users Fairly, Facebook Must Commit to Transparency."

At a private meeting a few days later, representatives of the board pressed the company on why it had, just a few months earlier, called the 6 million users protected by XCheck a "small number" and failed to address basic questions about the program. The two sides discussed how to address the inadequacies of Facebook's answer.

A month later, in October 2021, the board would announce in a quarterly report that it was conducting a review of XCheck at Facebook's request. The company conceded that it should not have told the board that XCheck applied to only a "small number" of decisions, but defended itself on the grounds that, for Facebook, 6 million users wasn't all that many.

The board chided Facebook for being "not fully forthcoming" and wrote that it had received assurance from the company that its answers would include more context "from now on."

And that was that. The following year, the board provided a number of pragmatic suggestions to improve the program and address some of its inequities, and Facebook agreed to accept many. With that, the Oversight Board's members, a collection of accomplished people who received a six-figure salary for a part-time job, would return to their core duty of adjudicating moderation calls on specific social media posts.

"There wasn't ever a 'How dare you!'" marveled the person familiar with Facebook's preparations for the Oversight Board's theoretical revolt.

As the *Wall Street Journal* team and I worked through the documents, Haugen and I remained in regular contact. We would talk every few days about something new one of us had spotted, acronyms we couldn't suss out (I eventually put together a glossary of more than three hundred terms), and how everything fit together.

She'd signed on with Whistleblower Aid, a DC-based legal non-profit that described itself as helping "patriotic government employees and brave, private-sector workers report and publicize their concerns—safely, lawfully, and responsibly."

The lawyers were never thrilled that Haugen had been working with a reporter directly. She passed on a request from them to be granted foreknowledge and approval of our publication plans. I owed Haugen a lot and was keenly aware of my responsibility to keep her identity confidential at all costs for as long as she wished, and to do my best to shield her from the fallout of our work. But no matter how grateful I was or how much I personally liked her, my job required that I serve the *Journal*'s readers, not assist her work as an advocate.

Haugen seemed to appreciate that distinction better than her lawyers, but there was tension nonetheless. Whistleblower Aid was telling her that she risked going to jail, and that if everything wasn't choreographed appropriately, the odds of that outcome grew. The precedents they were citing for this fear were odd, however, all involving federal employees or contractors with security clearances who had leaked classified government secrets. Haugen seemed skeptical of the risk, but neither of us was in a position to contradict the advice of counsel.

Haugen and her lawyers had decided to file a whistleblower complaint with the Securities and Exchange Commission. This meant that, at least in theory, she could not be sued for violating the non-disclosure agreement she had signed upon joining Facebook, since she had officially brought her concerns to the government, whether the SEC acted on them or not.

They also decided that Haugen would go public of her own accord, to undermine any possibility that Facebook could out her against her will, and arranged for an interview with *60 Minutes*. She taped her interview the week the *Journal's* series went live, and texted me a photo of her face displayed at different angles on a row of monitors. After several hours in the hands of a CBS-affiliated stylist, she had emerged as an almost unrecognizably prim version of herself.

"I look like a Texas news anchor," she wrote, adding that she wished her hair always looked that good.

Things were, however, getting more and more tense. Whistleblower Aid and some of a growing list of advisers on what I came to think of as Team Frances were pushing her to provide the documents to a wider range of outlets. They worried that the *Journal* having exclusive access would create resentment among other media and that, unable to get a piece of the story, those would attack Haugen.

This seemed to me both ridiculous and offensive. It was a way of thinking that treated media attention as a metric to be maxed out and Haugen's documents as currency. I argued that, after months of working with her, the *Journal* was the best positioned to tell the stories in the documents, and reminded her of how often it seemed like critical Facebook coverage had missed the mark. As Haugen herself had said to me, if the only effect of the leak was just that people hated Facebook a little more, then she would have failed.

I believed my arguments, and so did Haugen—at least enough to quash the proposal of a media-coordinated launch. But it was now clear that the *Wall Street Journal* had been drawn into a scrum of competing interests—everybody wanted a piece of Haugen's documents, and increasingly of Haugen herself.

—

Haugen and I always knew that the story about what Instagram was doing to the mental health of teenage girls would land hard. The gap between what the company had publicly said about the effect of its products on teenagers and the conclusions documented by its own researchers was huge. However much fun and value Instagram might provide most users, a subset of young women, namely those already in a vulnerable place, seemed to use the app compulsively and in ways that corroded their self-esteem. And Facebook knew it.

We also realized the story had the potential to cross party lines in the United States in a way that few things could. Facebook owed its ability to avoid heavy regulation from Washington to the fact that, even though both Democrats and Republicans hated Facebook, they could not agree why.

In the weeks before we pulled the trigger on the first story, both the *Journal* and Haugen's team had been in touch with the offices of Richard Blumenthal, the Democratic senator from Connecticut, and Marsha Blackburn, Republican senator from Tennessee, who were already sparring with the company over Instagram Kids and alleged harm to children.

Facebook seemed to understand the risk the story posed. The day after we approached the company for comment, revealing some of the research we had, Zuckerberg cleared a number of meetings from his calendar and the company began an internal review to assess the potential damage it would bring.

They engaged much more than they had on the XCheck story, offering a background interview with two of the platform's mental health researchers as well as a largely on-the-record conversation with Mosseri. "I think that we were late as a company to think through the downsides of connecting people at scale," Mosseri said. "I think there's way more upside than there is downside. But I do think there are downsides and we need to embrace that reality and do everything we can to address them as effectively as we can."

The research that the *Journal* had obtained showed Facebook's commitment to doing that, Mosseri said, and he was proud of the work. Social comparison, anxiety, and body image issues weren't specific to Instagram, but the company was doing the best it could to make sure it wasn't worsening them.

"I'm not saying that there aren't problems. I'm not saying that we shouldn't be further along than we are," Mosseri said. But all that the company could do now was progress as fast as it could. "We can't go back to 1960—like, there's going to be social media," he said.

The story ran on September 14, headlined "Facebook Knows Instagram Is Toxic for Teen Girls, Company Documents Show." Some of the voices cited were clinicians who treated young women for eating disorders, who had the sense that Instagram could be destabilizing for their patients, but the most prominent voices were those of the young women themselves. The story opened with Anastasia Vlasova, a high school tennis player who had developed an eating disorder after she began trying to emulate the preposterously restrictive diets that fitness influencers were promoting on Instagram.

Vlasova's experience was the worst-case scenario described in the Well-Being team's research: someone whose personal vulnerabilities had been exploited by personalized content. Accompanying her reported experiences were quotes plucked from teens who participated in Instagram's user experience research.

"After looking through photos on Instagram, 'I feel like I am too big and not pretty enough,' one girl told company researchers. 'It makes me feel insecure about my body even though I know I am skinny.'"

In its public response, the company struck a conciliatory tone. "We stand by this research," Karina Newton, Instagram's head of Public Policy, wrote in a post on Instagram's blog after the story went live. The *Journal,* she wrote, had focused on a "limited set of findings" and cast them in an excessively "negative light," but the findings were real. Instagram could give voice to the marginalized, connect people, and prompt social change, she wrote, "but we also know it can be a place where people have negative experiences, as the *Journal* called out today."

The public response was off the charts—the story dominated seemingly every newscast and online space.

Two days after it went live, Mosseri went on *Vox's Recode* podcast to talk about the story. He faulted the *Journal* for not giving the company enough credit for its efforts to understand and tackle the problem. "I'm biased, obviously, where I sit, my read on the article is largely one-sided, but I get where they're coming from and I don't want to quibble," he said. Then he said that social media's benefits come with a cost, likening the tradeoff to the one posed by cars.

"We know that more people die than would otherwise because of car accidents, but, by and large, cars create way more value in the world than they destroy," Mosseri said. "And I think social media is similar."

This was astonishingly respectful as far as pushback goes. The head of Instagram was acknowledging the potential for its product to cause harm.

Online there was a different reaction. "Instagram Chief Takes Heat for Bizarre Analogy Defending Social Media," wrote the *New York Post*, one of many publications that cited a list of spicy tweets from reporters and tech critics accusing him of being cavalier.

Mosseri took to Twitter himself to lament the number of reporters dunking on his "admittedly less than perfect" comparison between social media and cars. "Headline culture—which yes, I know, social media has contributed to—is exhausting," he wrote. Mosseri's frustration came a day after the *Journal* had run its third story in the series, on how Facebook's content recommendations had steered users toward increasingly angry, graphic, and upsetting content. The dynamic had been aggravated by the company's shift to "Meaningful Social Interactions," but it was just part of how content recommendation systems tended to work.

Twitter relied on a system like that, too. And, as Mosseri had just observed, the platform was an ideal environment to clown him.

"Instagram Boss Says Social Media Is Like Cars: People Are Going to Die," *Mashable* wrote in its roundup of the responses. The headline was perfect—and it proved Mosseri's point. Trying to talk about this stuff with a degree of sincerity hadn't gotten the company anywhere.

I watched the drama unfold and thought about the months I

had spent staring at Facebook documents that illustrated what was happening before my eyes. The same basic design that had been optimized to push people toward tweets insulting Mosseri was not dissimilar from that which promoted Facebook posts claiming an election had been stolen or Instagram accounts flaunting dangerously thin influencers.

The math behind the systems was complex, but the results were crude. They took human failings and encouraged them. Platforms promoted what would most likely get a rise out of users, and then content creators responded to those incentives. All of social media felt like poison in that moment.

I issued an apology to the editors for doubting Shark Week.

On a personal level, it had been a white-knuckle affair, with little sleep and a lot of stress. But the project, which we dubbed "The Facebook Files," laid out a picture of a company in crisis with dramatic flair. The *Wall Street Journal* had thrown every available resource into the documents, putting well over a dozen people on the project full-time. There was a separate project landing page, custom art, and a six-part podcast assembled via a partnership with Gimlet Media. We were promoting it with everything we had.

"They've gotta call the mercy rule at some point," tweeted Kevin Roose of the *New York Times,* summing up the intentional excess of our publishing schedule.

And nobody knew Haugen herself was coming.

Despite knowing so much about its users, Facebook appeared genuinely unable to identify which of its employees had siphoned secrets from scores of different places across its internal network. Not only had Facebook not contacted Haugen, but the company seemed confused about what documents the *Journal* had—something that a clean log of Haugen's activities should have mostly answered.

With no public whistleblower, the only entity for Facebook to strike back at was the *Journal.* Zuckerberg was upset that Facebook hadn't defended itself more stridently against the paper's reporting out of the gate. The CEO had made clear that he did not believe that apologizing was either warranted or useful. The Cambridge Analytica scandal had taught him, probably correctly, that corporate mea culpas wouldn't earn Facebook forgiveness, and longtime board members Peter Thiel and Marc Andreessen were now urging an aggressive pushback. The tenor of the company's response quickly changed, starting with a message from Nick Clegg, posted to the company blog the weekend after our stories ran.

Titled "What the Wall Street Journal Got Wrong," Clegg's post accused us of "cherry-picking" quotes and "deliberate mischaracterizations," while offering no specifics. He wrote that "we fundamentally reject this mischaracterization of our work and impugning of the company's motives."

Facebook couldn't pretend that we hadn't drawn blood on multiple fronts, however. XCheck was a politicized, self-interested mess. The company had acted on widespread human trafficking only when Apple's App Store stepped in as its surrogate conscience. Antivaccine activists had run circles around the defenses of a company with a passionately pro-vaccine CEO. But none of these things were as punishing as the material on Instagram and kids, and we weren't done on that front. Shortly after Shark Week, we asked Facebook for comment on a story about the company's plan to recruit its next generation of users.

Haugen had photographed some strategy documents showing that, with Snapchat and TikTok winning young users, Facebook was making a series of "big bets" on the company's future, focusing on the very young. Among the company's market research was a presentation titled "Exploring Playdates as a Growth Lever." If the company recruited enough users to Instagram Kids and Facebook Kids, it could nudge them toward Instagram when they came of age. As those users grew up, the company hoped, they would eventually turn to its namesake platform, where products like Groups and Marketplace would help Facebook become a "Life Coach for Adulting."

The plan required a lot to go right—starting with the successful launch of Instagram Kids. The idea of an Instagram for children was greeted with an outcry when it was first revealed by *BuzzFeed News* in March 2021. The company had shrugged off the criticism then, and in the wake of the teen mental health story it applied the same technique, dispatching hapless mid-level executives to answer questions like "Has Facebook quantified how many additional teens took their life because of your products?" from Congress.

But the ground was shifting. After first saying that it stood by the research that the *Journal* had obtained and acknowledging that the company hadn't found a way to remediate its own concerns about Instagram's effects on teen mental health, Facebook's head of Research rolled out a new position: actually, some of its own work sucked.

Describing her own staff's conclusions as mere "hypotheses," Pratiti Raychoudhury said the sample size for some of their work was too small to be meaningful and their stated conclusions were "not entirely accurate." Their warnings that Instagram contributed to negative social comparison and body image issues was "not supported by the studies."

The researcher flagellation did not turn the tide. One day before we published our next story on Instagram's child recruitment efforts, the company "paused" Instagram Kids. Although building the product was still "the right thing to do," Mosseri said, the company was indefinitely halting work "so we can get it right."

The company had just folded a business initiative it had, only a few months before, considered vital to its future. After that, its responses to the *Journal* got harsher, both externally and internally.

By the end of September, Haugen's anonymity was about done for. The *60 Minutes* interview was slated for October 3, a date that had been arranged after tense negotiations among her team, CBS, and the *Wall Street Journal*. That afternoon, reporters at the *Journal* gathered to watch the show. I had written a profile to launch at the same time.

"Meet @franceshaugen, who I've been calling 'Sean' for the past ten months," I wrote on Twitter with a link to the story. "Frances will be speaking for herself from here on out."

And so she did. Her appearance on *60 Minutes* was polished, and it received even more attention when Facebook suffered a six-hour global outage the next day. With the company unclear on what had happened, some media outlets began speculating about the possibility of sabotage. But, in keeping with corporate character, it was just an ill-timed technical fuckup. The company had accidentally deleted itself from the internet during routine maintenance, a mistake that could be remediated only by physically breaking into one of its own data centers.

Things were moving unbelievably quickly. Two days after the *60 Minutes* interview aired, Haugen was sitting before Congress, pronouncing that Facebook had lost its way. She walked the Senate Commerce Committee's subcommittee on consumer protection through the company's failings, most of which she traced back to Zuckerberg.

"Facebook wants you to believe that the problems we're talking about are unsolvable," Haugen told the committee, before laying out potential solutions, from reining in virality to rolling back engagement-based recommendation systems to rethinking platform design.

Haugen leaned in heavily to what Facebook had found about its effects on teenage girls, trying to break through to a geriatric body without a good grasp of the technical and conceptual bases for the proposals she was laying out.

Engagement-based ranking could take a user from "healthy recipes" toward pro-anorexia content at lightning speed, she warned. Underperforming and overhyped automated moderation systems meant that ads for "drug paraphernalia" were being served to teens. The Senate, she said, needed to act to protect "the safety of our children."

It was an odd act to watch from someone I had gotten to know so well. But Haugen was doing what was necessary to get her message across, playing up her Midwestern roots and smoothing over anything that might trip up the committee members or her TV audience. If the technical details of ranking systems and classifiers suffered a bit, it was in the service of making her point.

Jerry Moran, the Republican senator from Kansas, turned to Democratic senator Blumenthal and said, "The conversation so far reminds me that you and I ought to resolve our differences and

introduce legislation." Blumenthal agreed: "Our differences are very
minor, or they seem very minor, in the face of the revelations that
we've now seen."

Facebook's Communications team was apoplectic. Minutes after
the hearing ended, a spokesperson issued a testy statement noting
that Haugen had no access to Facebook's C-suite and did not work
on "the subject matter in question."

Accompanying the snub was a dare. It had been twenty-five years
since Congress last wrote the rules of the internet, and it was high
time they revisited them. "Instead of expecting the industry to make
societal decisions that belong to legislators, it is time for Congress to
act," the spokesperson said.

Facebook definitely did not mean that. In an effort to forestall
action, the company soon began an effort to discredit Haugen as a
whistleblower. Ironically enough, the effort benefited strongly from
the fanning of conspiracy theories.

Even before Haugen had gone public, rumors were spreading
on Capitol Hill that she was working with Fusion GPS, the strategic intelligence firm best known for producing the Steele dossier,
a collection of salacious but unfounded allegations that Russians
had compromised Donald Trump. We learned of the rumor when
a Republican congressional staffer who'd taken an interest in Instagram's mental health effects called in a panic, worried that they had
unwittingly tarred themselves by association with the notorious
firm.

My colleagues and I couldn't track down the source of the rumor,
but some efforts to sow doubt about Haugen among conservatives
were coming from Facebook itself. No sooner had she come forward than reporters from other news outlets began reaching out to
ask what I knew about the documents' origin—contacts at Facebook
had insinuated that there was more to the story than met the eye.
Did I think there was any possibility they might be right?

I hadn't appreciated how heavily the company was pushing that
line until the whisper campaign reached me. A Facebook Communications staffer called with a question: Was I truly so naive as to
believe that Haugen was just a disillusioned employee acting out

of conscience? Her information collection was too thorough, her congressional testimony too collected, her tale of joining Facebook out of idealism too tidy. The truth would out, this person assured me: Haugen was a professional plant, working on behalf of a deep-pocketed adversary.

I put everything I had into hearing out this theory without laughing. I'd been there for her solo document collection, of course. I wasn't surprised by her poise in front of Congress, given that, back at her parents' house in Ames, Iowa, there was a shelf of trophies from her championship-level collegiate debating career. And her backstory—especially that bit about going to work at Facebook as a result of a friend's descent into online white nationalism—had seemed a little too good to be true even to me, until I confirmed the story with the guy, who had snapped out of his racist phase, and Haugen pulled up the 2019 Gmail correspondence in which she told a Facebook recruiter why she'd wanted to work on misinformation.

Mainstream press never really took the alleged plot against Facebook seriously. But the company did find defenders in conservative media outlets—especially outlets that depended on Facebook.

Though Dan Bongino, a former Secret Service agent turned talk show host, and Ben Shapiro, a pundit who runs the *Daily Wire*, regularly competed with each other for the top slot on Facebook's engagement charts, on Haugen they saw eye to eye. Both men branded her a stalking horse for liberal censorship, citing her past political donations and current representation by a Democratic PR firm as proof that neither she nor the information in her documents deserved serious hearing.

"I am begging conservatives out there: please do not fall in this trap and work with this woman," Bongino announced in a video viewed on Facebook 1.2 million times. "This was all planned, this was all a plot." The *Daily Wire* pronounced Haugen a "leftist activist," and Shapiro declared that any legislation written in response to the leaked files would result in "tyranny."

In subsequent congressional hearings, partisan splits reemerged. Colleagues at the *Journal* later discovered that Facebook's Washington, DC, staff had played a role in the efforts to tar Haugen, as

well as our reporting, with sinister, partisan motives on Capitol Hill,
though we never found evidence that the company had tried to rally
conservative publishers to its defense. There may not have been a
need to. Partisan publishers like Bongino and Shapiro framed *everything* as having sinister motives—it was what had made them so
successful on Facebook.

However polished Haugen's debut as an advocate may have
looked, behind the scenes things were getting messy. On the day
before her congressional testimony, Whistleblower Aid had given
60 Minutes permission to post all of her SEC filings—documents
that happened to preview stories we had yet to publish. (Whistleblower Aid later said the publication of the documents had been the
result of a miscommunication.) The firm told Haugen that the move
had been a misunderstanding, but I was enraged. I called Haugen
and accused Whistleblower Aid of front-running our future work
and then lying to both of us about it. The firm was not just duplici-
tous but incompetent, I fumed. What PR wizards thought it was a
good idea to dump hot documents via the website of a weekly TV
show that ran the day before?

Haugen later told me she'd taken my call on a street corner in DC
while being tailed by two friends. As soon as she got off the phone,
she cried. Her own circle of advisers already had its internal disputes,
and now the *Journal* and her lawyers were pulling in different direc-
tions. Tensions flared again a few days later when Whistleblower
Aid and her public relations advisers told us they were going to con-
vene a consortium of American news outlets that would receive the
documents we already possessed from Congress.

With the *Journal* still working on a half-dozen additional stories, the idea wasn't attractive. Haugen and I had previously talked
over her desire to get the documents into the hands of international
news organizations, but the entities that her team had in mind were
other major American media outlets. At Haugen's urging, the *Journal* agreed to join a conference call to talk it over with prospective
participants, only to realize there was no thought of collaboration—
just an embargo date after which everyone would publish their own
takes on the material. Brad and I left the call, and the remaining

reporters conferred and decided to rebrand the corpus of work gathered in Puerto Rico "The Facebook Papers."

If the *Journal* didn't like this arrangement—and we absolutely loathed it—we had ourselves to blame. Haugen had long ago given us permission to publish any and all of the documents we received, and we had initially discussed making a substantial archive of them available when we published. But we hadn't done so for a variety of reasons, ranging from potential legal exposure to the protestations of the *Journal*'s art department against presenting readers with 20,000-plus blurry, off-kilter screenshots. When "The Facebook Files" stories ran, they were accompanied only by brief, heavily tidied snippets.

This was probably a mistake. I had come to see that the documents were a Rosetta stone for understanding how the design of social media platforms altered how users interacted and the ways that could flow across a platform. My colleagues and I had surfaced the material that we considered the most compelling and impactful for the *Journal*'s readers, but there were other audiences—academics, regulators, social media startups—who might be able to put what Facebook knew to use. No matter how many words we wrote, it would be arrogance to consider our work definitive. But the role of releasing raw source documents simply wasn't something that the *Journal* wanted to do, and so we didn't.

The release of documents to the consortium was chaotic. Promises by Haugen's team to hold back screenshots relevant to the *Journal*'s remaining stories didn't come to pass. Whoever was in charge of redacting the names of Facebook staff and its contacts overseas—an effort that Haugen and I had agreed was essential to head off safety and privacy concerns—was patchy. Multiple people inside Facebook, including those working on integrity matters, later told me that the broad distribution of poorly redacted documents both upset their authors and subjected civil society groups overseas to harassment. (Whistleblower Aid later acknowledged problems with the rollout of the documents, but said it had prioritized getting out information that it believed bolstered Haugen's whistleblower status.) Though I could never connect those assertions with specific harms, I considered the redaction failures an unforced error—and I let Haugen know. She told me she was doing the best she could and that I was becoming a jerk. We probably both had valid points. I later apologized and she later dropped Whistleblower Aid for another law firm.

If I hadn't been so pissed about it all, the next couple of weeks would have been funny. Armed with 22,000 pages of frequently technical screenshots, at least twenty different news outlets were simultaneously approaching Facebook's Communications team for comment on multiple stories at once.

"Not pumped to have discovered a level of workload that surpasses 'last week of a political campaign,'" tweeted Drew Pusateri, a Facebook Communications staffer in Washington.

Some stories trickled out before the consortium's October 25 embargo lifted, but, beginning at midnight that day, twenty-four different news outlets published at least sixty-five stories, some of which I wished we had written and others pure retread. The sheer volume imbued the release with a sense of shock and awe, but watching Facebook carpet-bombed with its own documents left me uneasy. I had flattered myself by thinking that Haugen's documents and the *Journal*'s work could force a reckoning for a company that everyone mistrusted but few understood. I was left struck by the feeling that maybe all it added up to was a lot of bad press.

In an unfortunate bit of timing, the consortium's embargo date coincided with Facebook's earnings release. Shortly after 1:00 p.m., the company announced healthy usage and a $9 billion quarterly profit. "Facebook shares rise as investors focus on earnings beat and look past whistleblower document dump," wrote CNBC.

Just three days later, Facebook made another announcement: the company was now named Meta. Some commentators jeered that the rebrand was the result of "The Facebook Files," but that wasn't true. Zuckerberg had proposed the change in early 2021, and Facebook's lawyers had been quietly negotiating to acquire Meta-related trademark rights for months. In one instance, a two-person startup guessed that Facebook was behind the effort and refused to consider relinquishing their fledgling company name for less than the preposterous price of $10 million. They got it. Zuckerberg wanted to move on, and money was no object.

"Our brand is so tightly linked to one product that it can't possibly represent everything we're doing today, let alone the future," the CEO announced on October 28. The company was pivoting to focus on the "metaverse," a vision of a digital future where people would live, work, and party in a kind of virtual reality environment. "From now on, we're going to be the metaverse first, not Facebook first."

The fact that the word "metaverse" was drawn from *Snow Crash* by Neal Stephenson—a 1992 sci-fi novel in which people don virtual reality headsets to escape a societal collapse so profound that corporate franchises are the main source of authority—was no deterrent.

There were doubters inside the company. Virtual reality had been the future for decades, without ever becoming the present. Facebook had been investing heavily in it since acquiring headset manufacturer Oculus in 2014. Every year, Facebook promoted Oculus headsets at its annual developers conference, F8, and heralded the technology as being on the cusp of widespread adoption. Then it would do the same thing the following year, a pattern that became sufficiently repetitive that the company had to address the legacy of missed targets in 2019. Everyone who attended the talk was given a free headset.

But Zuckerberg was excited. He unveiled a new series of company values, part of an update to the rebranded company's "cultural operating system." Among those values was to "live in the future." Employees should consider themselves to be "Metamates."

"The metaverse is the next frontier in connecting people, just like social networking was when we got started," the CEO said. "Our hope is that within the next decade, the metaverse will reach a billion people."

Zuckerberg's announcement of this new digital world was accompanied by a sense inside the company that, after a grueling six weeks, the threat from Haugen's documents had passed.

A later analysis from the company's Brand Marketing division found that it took less than a week for usage of Facebook to rebound from the three-day period in which Haugen had gone public.

The worldwide media pile-on had certainly reached users, though. Even after the rebrand to Meta, 80 percent of American users said the company was harmful to political discourse. The ratio of users who believed its products harmed people's emotional health ran as high as two to one. A later marketing analysis concluded that, while users in poorer markets overseas seemed less concerned— perhaps because "the conversation around the usage of social media hasn't developed sufficiently"—there were no near-term prospects for repairing Facebook's brand. "In countries already in low trust in Facebook app, messaging is effectively ignored," the analysis declared.

Conventional company thinking had been that such a "brand tax" would cause users to drop the platform, and Integrity staffers had long relied on that argument to push for changes that could not be justified on engagement grounds. But 22,000 pages of documents exposing damning corporate secrets had not deterred users any more than past advertiser boycotts or moderation controversies. If Facebook's reputation wasn't driving off its users by now, what would?

Contrary to "our prior beliefs," a marketing presentation produced by the company declared, "the evidence makes the hypothesis of a direct causal relationship between sentiment and engagement unlikely."

The company couldn't afford to permanently ignore its reputation, the presentation warned. Its bad name would likely make introducing new products more difficult and draw further regulatory scrutiny. But there was no immediate threat to the company's core business. People would keep using Facebook and Instagram, no matter what they thought of the company that operated them.

19

For external purposes, Meta was done with Haugen and "The Facebook Files." Internally, the fallout was just beginning.

Zuckerberg had initially delegated the job of responding to our stories to his deputies, dispatching them to Washington to deal with the pushback effort. But after Haugen's first appearance in front of Congress, the CEO turned his attention to another important audience: the one in Menlo Park.

"I'm sure many of you have found the recent coverage hard to read because it just doesn't reflect the company we know," he wrote in a note to employees that was also shared on Facebook. The allegations didn't even make sense, he wrote: "I don't know any tech company that sets out to build products that make people angry or depressed."

Zuckerberg said he worried the leaks would discourage the tech industry at large from honestly assessing their products' impact on the world, in order to avoid the risk that internal research might be used against them. But he assured his employees that their company's internal research efforts would stand strong. "Even though it might be easier for us to follow that path, we're going to keep doing research because it's the right thing to do," he wrote.

By the time Zuckerberg made that pledge, research documents were already disappearing from the company's internal systems. Had a curious employee wanted to double-check Zuckerberg's claims about the company's polarization work, for example, they would have found that key research and experimentation data had become inaccessible.

The crackdown had begun.

Like the airbrushing of purged Soviet leaders out of official photos, for the most part the removal of information from open internal forums went unremarked. But occasionally the disappearances were just too ironic to ignore. A multi-thousand-employee forum devoted to content discovery was set to "secret" status, and Cox's 2017 "Stop the Line" memo—the one in which he urged employees to unilaterally act when they saw the company falling short of its obligations—was restricted, too.

"Nick Clegg has been telling researchers on Central Integrity not to share research and experiment writeups on Workplace," one company source wrote to me. "It seems that the lesson Facebook learned from the leak is knowledge is a liability."

Haugen and I had known this response would come. Unwelcome discoveries and leaks had been dimming the future of self-critical research long before she had raided the company's archives, but now it was getting darker fast.

One of the first casualties was the content of Facebook's annual Research Day, an internal celebration of research on the frontiers of online social science. Carried out in the style of a daylong TED Talk series, the event was slated for later that October. One of the presentations was due to tackle what Facebook called "Conversational Motifs," patterns in back-and-forth exchanges that could be used to determine whether the conversation taking place was healthy. Mechanically parsing the difference between a respectful exchange of views and a conversation that ends with "Fuck me? No, fuck YOU!!!" showed promise, allowing Facebook to determine, among other things, which users were human flamethrowers.

This was a feel-good development, free of any implication of callousness, neglect, or misconduct. But with 20,000 pages of work product already on the loose, Facebook wasn't taking any chances. Facebook's Communications team insisted that Conversational Motifs work be yanked from the event's lineup.

At an all-hands forum for user experience researchers in October, Pratiti Raychoudhury, the head of Research, told her staff that they had gotten sloppy and too assertive about their opinions. Joining by video, Cox concurred. The teen mental health story, he said, showed that well-intentioned work would be distorted by journalists. Going forward, he said, the company would have to keep a tighter rein on internal discussions.

The nature of those new strictures became clear a few weeks later in a November 1 memo to Integrity researchers titled "On Narrative Excellence." Written by Raychoudhury and twenty-three other managers and executives, the memo stated that Meta was committed to "work that is honest and critical of existing processes," but such work was "vulnerable to misinterpretation" and needed to be more "effective."

What followed was a lengthy instructional guide to self-censorship.

Before posting something, researchers should "pressure-test" the work by anticipating any concerns from the company's Policy, Communications, or Legal departments; explicitly note that anything but finished reports were preliminary; and append reasons why they might be wrong. Research best practices included limiting the audience for the material to as few people as possible, not attempting to link problems on the platform to offline harm, and avoiding assertions that the company had a duty to take action. Under no circumstances, a "companion guide" warned, should researchers express a belief that the company was violating any law.

The guide contained some Kafkaesque mandates. One memo required researchers to seek special approval before delving into anything on a list of topics requiring "mandatory oversight"—even as a manager acknowledged that the company did not maintain such a list.

The "Narrative Excellence" memo and its accompanying notes and charts were a guide to producing documents that reporters like me wouldn't be excited to see. Unfortunately, as a few bold user experience researchers noted in the replies, achieving Narrative

Excellence was all but incompatible with succeeding at their jobs. Writing things that were "safer to be leaked" meant writing things that would have less impact.

None of the two dozen managers listed on the note responded.

Some projects were canceled solely because their names were controversial. Facebook's Legal department shut down work on the company's "Good for the World" classifier—a predictor of whether a user would consider a post to be societally positive—because of the implication that Facebook was recommending content that was not.

The company's new approach was best summed up in a series of presentations accompanying yet another shake-up announced in mid-2022. All of the company's Integrity and societally focused teams would report into a new structure with a mission to "amplify the good that happens on Meta's technology platforms." That structure would, in turn, support efforts on the Facebook and Instagram apps to "increase awareness of Meta's positive impact on the world" and ultimately "win hearts and shift perceptions."

The entirety of Facebook's staff working on integrity and societal issues was now literally reporting to Marketing, and the effects weren't subtle. Social scientists had to seek approval not just to conduct research that touched on politics, climate change, bias, health, or user well-being, but even to *propose* studying those subjects or summarizing their past work.

With few exceptions, even employees who had thrown bombs internally had traditionally abstained from criticizing Facebook publicly, even after they left the company. One explanation was fear of the nondisclosure agreement all employees were forced to sign, a legal document stating that leaks caused "irreparable harm" to the company, for which it could seek "extraordinary relief in court."

Haugen hadn't just broken that pledge, she had desecrated it. Facebook had pointedly refused to tell Congress that it wouldn't take action against her, but, at the same time, it was taking no action against her. With such a broad swath of company research already being discussed by so many people outside the company, some employees thought, what was the harm in chiming in?

"I've been impressed about how much more open many ppl are

talking publicly," one source texted me. "Like I am not going to 'pull a Frances' so mouthing off on Twitter is no big deal."

Even Chakrabarti spoke out. After Civic had been disbanded by the company, Chakrabarti had gone on parental leave. When he returned, he spent a few more months considering taking on a new role, but nothing quite fit. In early September, just weeks before we began publishing "The Facebook Files," he announced he was leaving the company. For nine months, I had been trying to get in touch with him—by email, Twitter, LinkedIn, text messages, any way I could—to no avail.

He broke his silence by responding to our stories on Twitter. He would tweet a story and write a thread annotating it with his thoughts. "Fundamentally, the trust & safety solution space is far broader when platforms prioritize reduction of user harm over short-term reputational concerns," he tweeted about our XCheck story. "When integrity teams get frustrated with company leaders, it is usually due to a misalignment on which priority matters more."

The story on Mexican cartels and human trafficking, he tweeted, was "especially difficult for me to read because it touches on a topic that probably 'kept me awake' more than anything else when I was at FB. And that is, how can social networks operate responsibly in the global south?" In another tweet he wrote: "Though I really really deplore leaks, at least my friends better understand why my hair turned so gray over the last several years."

With nearly each story and major reaction from Facebook, Chakrabarti shared his thoughts. When Clegg declared it "plain false," in a tweet, that the company buried negative findings, Chakrabarti begged to differ. "In my view, it's actually not uncommon for FB researchers working on societal issues to feel ignored, or at least grossly underweighted, in decisions that have public policy or growth tradeoffs," he responded, suggesting Clegg talk with them.

Regular rebuttals from Chakrabarti, someone long publicly held out as an expert on integrity work, undercut company accusations that the *Journal* had misrepresented Facebook research. But they also accomplished something more profound. Chakrabarti didn't have a massive social media following, but it included hundreds of loyal former employees, many still at the company. When he started criticizing the company's pushback against our reporting, the message seemed clear: it was time to start speaking up.

Jeff Allen, the data scientist who had demonstrated how Macedonians and other bad-faith players manufactured virality, before leaving the company in late 2019, was already at work with Sahar Massachi, a former Civic engineer, on a nonprofit that could serve as a home for Facebook Integrity, Content Ranking, and Safety staffers in exile. Haugen had archived notable work from both men, and they had stayed in touch with colleagues who had left Facebook to work at rival platforms like Twitter, TikTok, YouTube, and Clubhouse. They were looking to create an organization that could offer reporters and policymakers the expertise necessary to explain the implications of content ranking, the limitations of moderation, and the stakes of platform design choices. Unfortunately, both of them were clueless about media relations, think tanks, and institutional donors.

That was where Katie Harbath, who had left Facebook in 2021, came in. She had plenty of media experience and was already working on various projects with the Atlantic Council and Bipartisan Policy Center. The idea of working with someone who had once worked for Joel Kaplan gave Allen and Massachi initial pause, but eventually they joined up.

The Integrity Institute went public in late October 2021.

In a story about its launch, tech reporter Issie Lapowsky gave a sense of the new organization's approach by quoting an excerpt from Allen's final Workplace post, a screed on how the company should approach low-quality publishers. "If you just want to write python scripts that scrape social media and anonymously regurgitate content into communities while siphoning off some monetary or influence reward for yourself . . . well you can fuck right off," the note read.

The Integrity Institute's formation did not go over well with Facebook. Several current employees told me managers had advised them against getting involved—though they did anyway. By early 2023, the nonprofit had the funding to support six staff members in addition to its founders, along with fifteen fellows and more than 160 members. The group has been hired by the European Commission's Digital Media Observatory, published analyses of social media transparency reports, and advised upstart social media platforms on design choices. Chakrabarti is listed as the first member of its board of advisers.

Other ex–Facebook staffers have taken on similarly public roles. Brandon Silverman resumed his push to make social media data public from the outside, testifying before Congress, helping draft U.S. legislation mandating social media transparency, and consulting on the Digital Services Act, a European Union effort to regulate social media content delivery, advertising, and moderation that goes into effect at the beginning of 2024.

Arturo Bejar stayed in the game, too, consulting with Meta's Oversight Board and providing advice to a coalition of state attorneys general investigating the effects of the company's products on young users. As I wrote this book, he regularly dropped by my house to talk over his ideas for how a social media platform like Facebook could be replumbed to cultivate the sort of social structures that encourage constructive interaction over mere attention seeking. An account Bejar ran on behalf of his friend, the composer Philip Glass, was being bombarded with inappropriate messages from a deranged fan, and Bejar said that I should look at how Meta handled such harassment, especially when it was directed at teenagers.

I took his suggestion. Only a few hours of poking around Instagram and a handful of phone calls were necessary to see that something had gone very wrong—the sort of people leaving vile comments on teenagers' posts weren't lone wolves. They were part of a large-scale pedophilic community fed by Instagram's recommendation systems.

Further reporting led to an initial three-thousand-word story headlined "Instagram Connects Vast Pedophile Network." Co-written with Katherine Blunt, the story detailed how Instagram's recommendation systems were helping to create a pedophilic community, matching users interested in underage sex content with each other and with accounts advertising "menus" of content for sale. Instagram's search bar actively suggested terms associated with child sexual exploitation, and even glancing contact with accounts with names like Incest Toddlers was enough to trigger Instagram to begin pushing users to connect with them.

Meta internally declared a child-safety-specific lockdown in response to the story, blocking thousands of hashtags, mass deleting accounts, altering content moderator training, and restricting its recommendations. Employees I trust assured me that senior executives were genuinely aghast at what the company had missed.

On my end, however, the horrors of the project were pervaded by a sense of déjà vu. The company hadn't kept an eye on its recommendation systems, allowing a malignant community to grow at astonishing rates. It had relied on automated detection systems that were imprecise and failed to adapt to obvious adversarial behavior. Glitchy software made everything worse: the company's internal review uncovered a bug that had regularly routed user reports of child exploitation to the trash.

Meta's failures on child safety looked a lot like its troubles elsewhere. In one final echo, Facebook Groups proved a cesspool. With help from Stanford researchers, we easily identified dozens of large-scale child porn trading clubs. Facebook was recommending groups with as many 70,000 users and names such as Little Girls, building a global community around a shared passion for child sex abuse.

As always, Meta's Integrity staffers knew. Ex-employees told me they'd warned of trouble and devised remediations for some of the problems years before, only to see them watered down on the grounds that they were too heavy-handed.

These efforts had been documented internally, of course. There were references and a few links to such work in the 22,000 screenshots Frances Haugen had grabbed. They're almost certainly still there, behind Facebook's walls, if perhaps less accessible to employees than they once were.

The story of Facebook's integrity work is, in many respects, the story of losses. A collection of data scientists, user experience researchers, and machine learning experts put years of their lives into trying to fix Facebook. Their work uncovered flaws and helped mitigate awful outcomes, but they failed to persuade the world's largest social media company to fundamentally reevaluate how it built and managed its products.

If Meta's former Integrity staffers can't claim victory, however, their employer didn't end up in the winner's circle either.

The week before we began publishing "The Facebook Files," the stock of soon-to-be Meta rose to $380 a share, an all-time high. While the *Journal*'s reporting certainly wasn't responsible for most of the corporate tumult that followed, some of the pitfalls of Zuckerberg's unilateral authority that we had documented certainly were.

Shortly after we began running the series, the *New York Times* wrote a story revealing the details of a public relations strategy that company leadership had adopted earlier in the year, a plan that boiled down to dismissing criticism and promoting Zuckerberg as a fun-loving visionary. To accentuate the point, Zuckerberg posted about the story on Facebook, complaining that the *Times* had incorrectly described him as riding an electric surfboard when the device was, in fact, a human-powered hydrofoil.

"I don't normally point out everything the media gets wrong, but it happens every single day," Zuckerberg wrote, joking about suing the *Times* for defamation. Rather than worry about criticism, the CEO wrote to a commenter, "I just decided to focus more on some of the awesome things we're building, and doing more fun things with my family and friends."

Internally, Zuckerberg flatly rejected a plan to publicly abandon the company's goal of building a preteen version of Instagram as a peace offering to Washington, ahead of congressional testimony by Mosseri. He ordered Meta's Communications team not to yield any ground to critics.

"When our work is being mischaracterized, we're not going to apologize," spokesperson Andy Stone told the *Journal*.

If any of the news coverage had drawn blood, Meta wasn't going to show it. Zuckerberg told the company's People Planning team to bring him an aggressive hiring target for 2022. When they brought him an unprecedentedly ambitious plan to bring on 40,000 new staffers that year, Zuckerberg took the one-page document—known as "the napkin"—and then passed it back with a handwritten instruction to hire 8,000 more.

"If we don't hit these targets it's game over," Recruiting VP Miranda Kalinowski told the managers on her staff. To handle the deluge of hiring, Meta brought on an additional 1,000 recruiters between the last quarter of 2021 and the first quarter of the following year.

Few of the new staffers would be slated to go into integrity work. Zuckerberg had declared that the company's existing products were no longer its future, and Haugen's document breach had solidified a sense that researchers and data scientists working on societal problems contained a potential corporate fifth column.

That assumption wasn't entirely wrong. Whatever the company's crackdown on integrity research had accomplished, the gains came at a cost. Not long after mandated "Narrative Excellence" took hold, a staffer who I had tried to contact months before got in touch. The person worked on a sensitive subject matter and had long been convinced they could do more good by working quietly rather than talking to a reporter. The company's new strictures on the work had broken them.

Within a few months, the *Journal* was again receiving a steady stream of documents from them and other employees—thousands of additional pages. Despite Meta's efforts to tame the unruly spread of information on Workplace, the material was still good.

One story from the new documents described how Zuckerberg had tried to get Facebook out of politics. Instead of following through on post-2020 efforts to address the platform's tendency to amplify outrage-based and sensationalistic political content, the CEO had asked that News Feed try demoting to oblivion anything Facebook classifiers deemed to involve politics, health, or social issues. Although Zuckerberg was willing to accept a hit to Facebook engagement in order to suppress what he had once praised as "the Fifth Estate," the effort ran aground. Political content escaped the platform's flawed classifiers, donations to charities tanked, and, worst of all, surveys showed that people still associated Facebook with politics. With experiments showing the modest payoff would come at "a high and inefficient cost," Zuckerberg personally pulled the plug.

Unable to escape the political conversation, the platform got back to work on what the Civic team and others had long pushed for: an end to recommending political and health content based solely on its ability to produce maximum engagement.

Another story based on freshly leaked documents demonstrated that Meta was trying to limit the distribution of recycled viral content in favor of entities that produced original material—a move that Silverman, Allen, and Facebook's Partnerships team had been pushing since late 2017, with limited success. What had changed was Meta's scramble to compete with TikTok by heavily promoting short-form videos from entities that users didn't follow. Freed from the final vestiges of user taste, Facebook and Instagram recommendation systems were going all in on bootlegged media, people fighting, and oversexualization. Fully 70 percent of the top twenty most-viewed posts on Facebook in the third quarter of 2021 met the company's formal definition of "regrettable," one document stated, with the remaining 30 percent being engagement bait. By early 2022, Meta had convened a "Content Quality War Room." Though Allen and Silverman, since departed from Facebook, were unimpressed when I gave them details of its work, it seemed like a start.

The most notable of the new documents weren't about the problems on Meta's platforms so much as the company's integrity work itself. For four years, Facebook had been making decisions premised on what Guy Rosen had termed "the usual growth vs. integrity tension." No matter whether the goal was fighting hate speech, slowing viral misinformation, or downranking unoriginal content, tests routinely showed that integrity ranking changes came at the cost of growth. Integrity-focused staffers would invariably argue that their proposals to clean up the platform would benefit Facebook in the long run, while their growth-focused colleagues would point to short-term experimental data and scoff.

In response, frustrated data scientists designed a highly unusual experiment, tracking the cumulative effect of integrity work against a control group known as the "minimum integrity holdout." By the end of 2021, the experiment had yielded enough data for scientists to write up a clear conclusion. Facebook had been screwing up how it measured the effects of integrity work for years.

All the watered-down integrity ranking changes they had shipped did, in fact, dent Facebook usage, the way the naysayers suggested—but only for six months. After that, and long after Facebook called off its experiments, the harm done to the company's engagement metrics ebbed into statistical insignificance. After about a year, the integrity work began producing a modest but statistically significant usage *gain*, especially among young and infrequent users. Network effects made the true size of the gain impossible to calculate, but it was likely larger than the fraction of a percent the data scientists could measure.

"To engender an ideal Facebook ecosystem—i.e., a highly engaging ecosystem with high integrity—we should invest in more (and more intense) integrity products," the note concluded.

The initial response to the research was skepticism. Because of what the note acknowledged were "major implications for Facebook," managers assigned several senior data scientists to probe the work for computational errors and system glitches that might have tainted the results.

But the gains were real. News Feed integrity work had boosted growth for years in ways the company hadn't understood.

The good news of this discovery was that Facebook could likely boost engagement by investing in stronger News Feed integrity ranking. The bad news was that, by blocking or watering down past integrity work, the company had likely been polluting its platform to the detriment of both users and itself.

The embarrassing implications held up approval to share the research for months. Managers eventually did sign off on circulating it among Integrity staffers but blocked its wider distribution because, two employees familiar with the matter said, the findings were considered too confrontational. Efforts to quantify the growth benefits of integrity ranking work continued quietly, with additional research crediting it with boosting daily usage among new users by 0.9 percent, a huge win on Facebook's most hallowed metric. People familiar with the work are hopeful that it will eventually prompt Facebook to adopt stronger integrity measures. After all, Zuckerberg isn't one to leave usage gains on the table.

As for Meta's chief executive himself, his effort to focus more on "the awesome things we're building" soon met with complications. Embracing remote work, Zuckerberg had just begun a three-month stint at his ranch in Montana in the spring of 2022 when Facebook's user growth fell for the first time, devastating Meta's stock and chilling the torrid hiring push he had ordered. By June of that year, the CEO told a company-wide town hall, "Realistically, there are probably a bunch of people at the company who shouldn't be here." The company's first-ever mass layoffs—11,000 people—came in November, and its second-ever mass layoffs began four months later. Integrity staff and user experience researchers got hit, hard.

Nor does the new digital world that Zuckerberg rechristened the company for appear as near as he suggested. The 2022 brand sentiment presentation the *Journal* obtained was prescient in its prediction that Meta's reputation would be an impediment when launching new products. Early market research noted that potential buyers of augmented reality glasses cited the social unacceptability of wearing company-branded glasses "without prompting."

Beyond the image problem, the company has also struggled to build the metaverse. Documents the *Journal* obtained shortly before the one-year anniversary of Facebook's rebranding showed that its virtual reality app Horizon Worlds was filled with bugs, but few users. Daily visitation had fallen to less than 200,000. Rather than focusing on social media apps with a combined 3.5 billion daily active users, Meta's CEO spent much of 2022 enthralled with one that had a user base a little smaller than the population of Sioux Falls, South Dakota.

It's too early to write off the metaverse, of course. A decade could prove Zuckerberg a visionary. But enthusiasm for that vision has waned, even inside Meta. Employee surveys consistently find low confidence in the company's leadership. In one memo that leaked,

Metaverse VP Vishal Shah scolded employees for failing to spend time in the metaverse themselves.

"The simple truth is, if we don't love it, how can we expect our users to love it?" the executive asked. A later Workplace post seen by the *Journal* invited employees to come learn about the company's HR benefits in the metaverse, an event that would count toward "weekly required headset time."

By the spring of 2023, Zuckerberg was backing off his plan for company-wide remote work that he had announced early in the pandemic. Meta employees were going to need to spend more time in the office, and Facebook and Instagram were going to need to keep performing. The company's social media platforms weren't the future of technology, but they weren't yet its past.

ACKNOWLEDGMENTS

This book is for my parents and my wife, Camas. While I never doubted that "The Facebook Files" could yield a book, my ability to write one was a bigger leap.

Beyond family, the first thank-you should go to Brad Reagan at the *Journal,* who hired me, put up with me, and oversaw the reporting that went into a very weird and all-consuming project. In a similar vein, I'm also grateful to the San Francisco bureau chief, Jason Dean, and my "Facebook Files" coauthors Keach, Georgia, Sam, Justin, Newley, Emily, and Deepa.

I'm grateful to my agent, Eric Lupfer, for helping me figure out a book proposal in which the protagonists are data scientists. After making the dubious decision to commission a book titled *Facebook Files TBD,* Doubleday editor Yaniv Soha was a pleasure to work for/with, helping shape the book and giving me the runway needed to make *Broken Code* more than a rehash of past reporting.

Miriam Elder was indispensable as an outside editor. Stepping in after my first editor washed his hands of me for entirely understandable reasons, she helped me identify a viable structure for a story in which things often happened simultaneously and then hammered my reporting into a better draft than I thought possible.

I found fact-checker Sean Lavery toward the end of the process. He is brilliant, meticulous, and profoundly likable. While I've yet to meet him in person, he's a friend.

Rebecca Crootof, Eliot Brown, and Emily Sachs all provided helpful suggestions on the book's first draft. Hannah Kekst helped with the endnotes.

Finally, there are many Facebook/Meta employees who gave me both their trust and their time. Many are named in the book, but far more can't be. One in particular provided almost all the source material for events cited in 2022, but dozens of others influenced my thinking and corroborated events. They spoke to me in the hope that the book would accurately record their work and show how social media could be different. Thank you.

NOTES

This book relies on three sources of information to tell the story of how Meta Platforms Inc., formerly Facebook, came to understand the ways in which it was changing how billions of people communicate and interact on its platforms. The first is more than 25,000 pages of internal company work product, provided to the *Wall Street Journal* by Frances Haugen as well as other current and former employees. To protect the privacy and safety of the staff-level researchers, engineers, and data scientists who authored these documents, their names have generally been excluded. Senior managers and executives are cited when their identities are relevant.

Beyond contemporaneous documents, this book relies on interviews with more than sixty people, who provided context regarding these records as well as recollections regarding their work at Facebook. Many of these sources requested anonymity to avoid antagonizing their current or former employer. Although some of the prominently cited figures in this book chose to speak with me, others did not. In those instances, quotes are derived from people who witnessed the events in question or were directly informed of them thereafter.

The company also authorized background interviews with a range of its executives and later approved some information from those conversations to be brought on record. Those instances are noted in the text. Additionally, the company's Communications staff was given an opportunity to challenge both specific events cited herein as well as the interpretations of its internal work product.

Works previously published by the *Wall Street Journal* or other news outlets are cited in the notes.

CHAPTER 2

14 A tough-guy mayor: Charlie Campbell, "The Philippine Election Front-Runner Calls His Daughter a 'Drama Queen' for Saying She Was Raped," *Time,* April 20, 2016.

17 He had gotten into Trumpworld: Peter Elkind with Doris Burke, "The Myths of the 'Genius' Behind Trump's Reelection Campaign," ProPublica, September 11, 2019.

19 "another Twitter moment": Mike Isaac and Sydney Ember, "For Election Day Influence, Twitter Ruled Social Media," *New York Times,* November 8, 2016.

20 "Facebook and Twitter were the reason": Issie Lapowsky, "Here's How Facebook *Actually* Won Trump the Presidency," *Wired,* November 15, 2016.

21 "After the election, many people are asking": Mark Zuckerberg, "I want to share some thoughts on Facebook and the election," Facebook, November 12, 2016, www.facebook.com/zuck/posts.

21 One week after the election: Craig Silverman, "This Analysis Shows How Viral Fake Election News Outperformed Real News on Facebook," *BuzzFeed News,* November 16, 2016.

CHAPTER 3

25 A two-month sprint: Antonio García Martínez, *Chaos Monkeys: Obscene Fortune and Random Failure in Silicon Valley* (New York: HarperCollins, 2016).

26 According to a later history: Author name redacted, "Why FB App Took on Sessions as a Top-Line Metric for 2019," Facebook memo, October 13, 2020.

26 As Zuckerberg explained: Jerrold Nadler and David N. Cicilline, "Investigation of Competition in Digital Markets," U.S. House of Representatives, Subcommittee on Antitrust, Commercial, and Administrative Law of the Committee on the Judiciary, October 2020, www.govinfo.gov/content.

27 During one six-month period: Ben Maurer, "Fail at Scale: Reliability in the Face of Rapid Change," ACM Queue, October 27, 2015, https://queue.acm.org.

28 In 2021, when Mike Schroepfer: Mike Schroepfer, "What's Slowing You Down?," Facebook Workplace memo, April 26, 2021.

28 "The best antidote to bad speech": Michal Addady, "Facebook Enlists Users' Help to Fight Terror," *Fortune,* January 21, 2016.

CHAPTER 4

65 Such distrust was hard: Hannah Kuchler, "Facebook Employees Overwhelmingly Back Democrats," *Financial Times*, October 30, 2018.

CHAPTER 5

72 "Harm comes from situations": Author name redacted, "First Do No Harm," Facebook memo, February 2020.

73 Those hurdles meant that integrity: "First Do No Harm."

75 Citing the plummeting traffic: Sheera Frenkel, Nicholas Casey, and Paul Mozur, "In Some Countries, Facebook's Fiddling Has Magnified Fake News," *New York Times*, January 14, 2018.

77 "There's too much sensationalism": Mark Zuckerberg, "Continuing our focus for 2018 to make sure the time we all spend on Facebook is time well spent," Facebook, January 19, 2018, www.facebook.com/zuck.

79 "Based on this, we're making": Mark Zuckerberg, "One of our big focus areas for 2018 is making sure the time we all spend on Facebook is time well spent," Facebook, January 11, 2018, www.facebook.com/zuck.

79 The goal became constant: Author name redacted, "Why FB App Took on Sessions as a Top-Line Metric for 2019," Facebook memo, October 13, 2020.

81 Under the new MSI system: Author name redacted, "Deriving MSI Weight," Facebook internal research, December 2017.

84 *Slate* noted that the most-shared: Will Oremus, "How a 119-Word Local Crime Brief Became Facebook's Most-Shared Story of 2019," *Slate*, March 29, 2019.

84 Facebook's after-the-fact research: Common Ground team, "Case Study: (Controlling for Publisher) Posts with Negatively Charged Comment Threads Fare Better in Feed," Facebook internal case study, November 2018.

84 The effect was more than: Author name redacted, "Political Party Response to '18 Algorithm Change," Facebook internal report, April 2019.

84 "They have learnt that harsh": "Political Party Response."

85 Extremist parties proudly told: "Political Party Response."

85 To compete, moderate parties: "Political Party Response."

85 To the researcher this sounded: "Political Party Response."

85 Later research conducted in Asia: Author name redacted, "Overview of Polarization Research (Literature Review)," Facebook report, June 2020.

85 more serious topics: "Social Media & the January 6th Attack on the U.S. Capitol: Summary of Investigative Findings," U.S. House Select Committee to Investigate the January 6th Attack on the United States Capitol, leaked draft document, January 2023, accessible at https://www.washingtonpost .com/documents/5bfed332-d350-47c0-8562-0137a4435c68.pdf.

CHAPTER 6

93 "The mission of Facebook": Jeff Allen, "Making Web Platforms More Resilient Against Disinformation," Facebook memo, February 2019.

93 "This is not normal": Jeff Allen, "How Communities Are Exploited on Our Platforms: A Final Look at the 'Troll Farm Pages,'" Facebook memo, October 2019.

94 One of Allen's colleagues: Allen, "How Communities Are Exploited on Our Platforms."

94 "What's the easiest": Jeff Allen, "Making Web Platforms Resilient Against Disinformation," Facebook presentation, February 20, 2019.

94 "We should remove content": Jeff Allen, "Exploiting Communities: Using Qualitative, Objective Signals to Identify Content That Exploits Communities, Like BuzzFeed Did," Facebook memo, July 2018.

96 A 2019 screening: Jeff Allen, "A Signal to Demote Pages Manufacturing Virality with Unoriginal Content," Facebook memo, July 2019.

101 When Media Matters, a liberal: "2019 Guide to Publishing on Facebook," NewsWhip, March 10, 2019, https://go.newswhip.com.

102 Distinguishing himself from publishers: Michelle Rennex, "21 Things That Almost All White People Are Guilty of Saying," BuzzFeed, September 19, 2018.

CHAPTER 7

113 Once Facebook cloned: Deepa Seetharaman, "Facebook's Onavo Gives Social-Media Firm Inside Peek at Rivals' Users," *Wall Street Journal,* August 13, 2017.

117 A UN report declared: Human Rights Council, "Report of the Detailed Findings of the Independent International Fact-Finding Mission on Myanmar," United Nations report, A/HRC/39/CRP.2, September 17, 2018, www.ohchr.org.

118 "has become a means": BSR, "Human Rights Impact Assessment: Facebook in Myanmar," October 2018.

CHAPTER 8

127 That wouldn't become clear until: Sophie Zhang, Facebook memo, January 2020.

127 "Should we not be": David Agranovich, "Should We Not Be Looking at Whether We Can Adjust?," Facebook Workplace comment, January 2020.

143 In a statement, the company: Nathaniel Gleicher, "Removing Coordinated Inauthentic Behavior and Spam from India and Pakistan," Meta, April 1, 2019, https://about.fb.com/news/2019.

146 "These examples are so blatant": Author name redacted, "Low Quality Civic Exports Targeting the US," Facebook report, August 2019.

CHAPTER 9

148 In July 2019, Jin wrote: Kang-Xing Jin, "Virality Reduction as an Integrity Strategy," Facebook Workplace memo, July 2019.

149 A few months after that exchange: "How to Evaluate Experiments for Visitation," Facebook guide, 2019.

151 "We cannot assume links": Public Policy team, "BWC GTM: Deprecating Sparing Sharing and Informed Engagement (Preliminary Proposal)," Facebook writeup, 2019.

152 A few months after his first: Kang-Xing Jin, "Defining Success in Addressing Integrity Harms, Starting with Defining Our Responsibilities," Facebook memo, October 2019.

158 "prioritizing PR risk over social harm": "Social Media & the January 6th Attack on the U.S. Capitol: Summary of Investigative Findings," U.S. House Select Committee to Investigate the January 6th Attack on the United States Capitol.

160 Though his name: Samidh Chakrabarti, "US 2020 Leadership Update," Facebook report, March 3, 2020.

CHAPTER 10

163 By the time of a 2018 report: Author name redacted, "State of Teens," Facebook Report, June 2018.

165 what researchers termed "risky behavior": Agam Bansal et al., "Selfies: A Boon or Bane?," *Journal of Family Medicine and Primary Care* 7, no. 4 (Summer 2018): 828–831, https://doi.org/10.4103/jfmpc.jfmpc_109_18.

165 By 2017, a British study: Royal Society for Public Health and the Young

Health Movement, "#StatusOfMind: Social Media and Young People's Mental Health and Wellbeing," May 2017, www.rsph.org.uk.

165 "I have no doubt that Instagram": Ian Russell interview by Angus Crawford, "Instagram 'Helped Kill My Daughter,'" BBC, January 22, 2019.

165 Later research—both inside and outside: Amanda MacMillan, "Why Instagram Is the Worst Social Media for Mental Health," *Time,* May 25, 2017.

166 Instead, the researchers wrote: Instagram Well-Being team, "Teen Mental Health Deep Dive," Facebook report, October 2019.

167 "They often feel 'addicted'": "Teen Mental Health Deep Dive."

168 "People see about 5%": Author names redacted, "Social Comparison: Topics, Celebrities, Like Counts, Selfies," Instagram report, January 28, 2021.

168 As Mosseri put it: Adrienne So, "Instagram Will Test Hiding 'Likes' in the US Starting Next Week," *Wired,* November 8, 2019.

CHAPTER 11

176 When Italy became the first: "Facebook Group Calls Soar 1,000% During Italy's Lockdown," BBC, March 25, 2020.

177 Facebook had been struggling: Tom Alison, "Why Hiring Is Hard Right Now," Facebook HR memo, 2019.

177 He also wrote to Anthony Fauci: Jason Leopold (@JasonLeopold), "NEW via my #FOIA: Email from Mark Zuckerberg to Anthony Fauci on March 15," Twitter, December 2, 2020, https://twitter.com/JasonLeopold.

178 "The Pandemic Is Giving": Sarah Frier, "The Pandemic Is Giving Zuckerberg a Shot at Making Amends," *Bloomberg Businessweek,* April 16, 2020.

178 A *New York Times* feature: Mike Isaac, Sheera Frenkel, and Cecilia Kang, "Now More Than Ever, Facebook Is a 'Mark Zuckerberg Production,'" *New York Times,* May 16, 2020.

178 On May 25, 83 percent: Facebook internal company dashboards, May 25, 2020.

178 In May, a data scientist: Author name redacted, "Facebook Creating a Big Echo Chamber for 'the Government and Public Health Officials Are Lying to Us' Narrative—Do We Care?," Facebook Workplace memo, May 2020.

178 An analysis showed these: "Facebook Creating a Big Echo Chamber."

179 "This is severely impacting": Author name redacted, "This Is Severely Impacting Public Health Attitudes," Facebook Workplace comment, May 2020.

180 Over on Facebook: "Hate Begets Hate; Violence Begets Violence," Facebook Classifier for Violence and Incitement, June 2020.

182 XCheck was used by more: Author name redacted, "Whitelist: Where We Are and Where We Want to Be," Facebook report, November 2019.

182 Later reviews would find: Author name redacted, "XCheck—Get Well Plan," Facebook review, June 2020.

182 An internal review of the incident: Author name redacted, "Mistake Prevention Incidents Investigation," Facebook report, 2021.

183 "Unlike the rest of our community": Author name redacted, "The 'Whitelist' Problem," Facebook review, 2019.

183 The company had sent: Avi Selk, "Facebook Told Two Women Their Pro-Trump Videos Were 'Unsafe,'" *Washington Post,* April 10, 2018.

184 According to documents first obtained: Craig Silverman and Ryan Mac, "Facebook Fired an Employee Who Collected Evidence of Right-Wing Pages Getting Preferential Treatment," *BuzzFeed News,* August 6, 2020.

185 "At the end of June 2": "Hate Begets Hate; Violence Begets Violence."

186 Zuckerberg decided instead to do nothing: Shirin Ghaffary, "Mark Zuckerberg on Leaked Audio: Trump's Looting and Shooting Reference 'Has No History of Being Read as a Dog Whistle,'" *Vox,* June 2, 2020.

186 Chakrabarti wrote a post: Samidh Chakrabarti, "Bending Our Platforms Toward Racial Justice," Facebook Workplace memo, June 2020.

187 The post received a decidedly cooler: "Social Media & the January 6th Attack on the U.S. Capitol: Summary of Investigative Findings," U.S. House Select Committee to Investigate the January 6th Attack on the United States Capitol, 53.

188 Hours after the Q&A: Shirin Ghaffary, "Read the Transcript of Mark Zuckerberg's Tense Meeting with Facebook Employees," *Vox,* June 3, 2020.

188 The company had repeatedly: Craig Silverman and Ryan Mac, "Facebook's Preferential Treatment of US Conservatives Puts Its Fact-Checking Program in Danger," *BuzzFeed News,* August 13, 2020.

189 "The language used by these": Author name redacted, "The Language Used by These Events and Groups Was Not Violating," Facebook Workplace memo, August 2020.

190 Chakrabati should never have: "Social Media & the January 6th Attack on the U.S. Capitol: Summary of Investigative Findings," U.S. House Select Committee to Investigate the January 6th Attack on the United States Capitol.

CHAPTER 12

191 The week before, the *Wall Street Journal:* Newley Purnell and Jeff Horwitz, "Facebook's Hate-Speech Rules Collide with Indian Politics," *Wall Street Journal,* August 14, 2020.

192 Facebook removed hate speech: Ajit Mohan and Monika Bickert, Facebook workplace memo, August 2020.

193 "Hindus, come out": Madan B. Lokur et al., "Uncertain Justice: A Citizens Committee Report on the North East Delhi Violence 2020," Constitutional Conduct Group, October 2022, https://constitutionalconduct.com.

195 Ahead of the Indian elections: Author name redacted, "An Indian Test User's Descent into a Sea of Polarizing, Nationalistic Messages," Facebook report, February 2019.

197 In 2019, the BBC: Owen Pinnell and Jess Kelly, "Slave Markets Found on Instagram and Other Apps," BBC, October 31, 2019.

197 "Removing our applications": Author name redacted, "Apple Escalation on Domestic Servitude—How We Made It Through This SEV," Facebook report, November 2019.

198 Two years later, in late 2021: Author name redacted, "Domestic Servitude: This Shouldn't Happen on FB and How We Can Fix It," Facebook report, February 2021.

198 One memo noted: Author name redacted, "Domestic Workers Awareness Project: A Preventative Approach," Facebook memo, 2020.

199 An internal review referred: Author name redacted, "Responsible Enforcement at Facebook: A Method for Surfacing Fairness Issues in Labeling," Facebook internal review, 2021.

200 "Most of our integrity systems": Tony Leach, Integrity Country Prioritization for 2021, December 10, 2020.

201 Events on the platform: Tessa Knight and Beth Alexion, "Influential Ethiopian Social Media Accounts Stoke Violence Along Ethnic Lines," Digital Forensic Research Lab (DFRLab), Medium, December 17, 2021.

201 His family is now suing: Alex Hern, "Meta Faces $1.6bn Lawsuit Over Facebook Posts Inciting Violence in Tigray War," *Guardian,* December 14, 2022.

202 Facebook told Reuters: James Pearson, "Exclusive: Facebook Agreed to Censor Posts After Vietnam Slowed Traffic," Reuters, April 21, 2020.

205 One was about Facebook's: Jeff Horwitz and Newley Purnell, "In India, Facebook Fears Crackdown on Hate Groups Could Backfire on Its Staff," *Wall Street Journal,* December 13, 2020.

205 Kiran was also: Jeff Horwitz and Newley Purnell, "Facebook Executive Supported India's Modi, Disparaged Opposition in Internal Messages," *Wall Street Journal,* August 30, 2020.

CHAPTER 13

207 A lone user could: Author name redacted, "Problematic Non Violating Narratives Is a Problem Archetype in Need of Novel Solutions," Facebook report, March 2021.

208 Emails later turned: "Social Media & the January 6th Attack on the U.S. Capitol: Summary of Investigative Findings," U.S. House Select Committee to Investigate the January 6th Attack on the United States Capitol, leaked draft document, January 2023, accessible at https://www.washingtonpost.com/documents/5bfed332-d350-47c0-8562 -0137a4435c68.pdf.

209 In a presentation to leadership: Author names redacted, "Dangerous Civic Groups," Facebook presentation, September 2020.

210 enforcement for "low-tier hate speech": "Social Media & the January 6th Attack on the U.S. Capitol."

212 "This launch is temporary": Author name redacted, "[LAUNCH] Using p(anger) to Reduce the Impact Angry Reactions Have on Engagement Ranking Levers," Facebook report, September 2020.

214 These gains—which a memo: Author name redacted, "[Launch] Reshare Depth Demotion—Myanmar," Facebook report, October 2020.

214 "We plan to roll": "[Launch] Reshare Depth Demotion—Myanmar."

215 In September, Nick Clegg had: Editorial board interview with Nick Clegg, "How Facebook Tries to Fight Being a Superspreader of Fake News on Voting and COVID-19," *USA Today,* September 23, 2020.

216 In an email later obtained: U.S. House Select Committee, "Social Media & the January 6th Attack on the U.S. Capitol."

217 And though its relations: Julie Bykowicz, Brody Mullins, and Emily Glazer, "Big-Name Democrats Say 'No Thanks' to Facebook's Top Lobbyist Job," *Wall Street Journal,* October 26, 2021.

219 Later analysis would confirm: Author name redacted, "Stop the Steal and Patriot Party: The Growth and Mitigation of an Adversarial Harmful Movement," Facebook report, March 2021.

220 Strengthened demotions of incitement to violence: "Social Media & the January 6th Attack on the U.S. Capitol."

CHAPTER 15

231   The Kremers' first attempt: Brandon Straka, interview with the U.S. House Select Committee to Investigate the January 6th Attack on the United States Capitol, 2021.

231   "I'm gonna do it": Curt Devine, Drew Griffin, and Zachary Cohen, "Alex Jones Allegedly Threatened to Throw Trump Rally Organizer Off a Stage," CNN, March 12, 2021.

232   Or as she put it in a text: U.S. House Select Committee to Investigate the January 6th Attack on the United States Capitol, "Hearing on the January 6th Investigation, Second Session," House of Representatives, 117th Congress, July 12, 2022, www.congress.gov.

235   Only later would the company: Author name redacted, "Stop the Steal and Patriot Party: The Growth and Mitigation of an Adversarial Harmful Movement," Facebook report, March 2021.

236   On December 17: Author name redacted, "Mission Control: Response HPM—Jan. 21," Facebook review, January 2021.

236   A later review of Facebook's: "Mission Control: Response HPM."

237   "Hang in there everyone": Mike Schroepfer, "Hang In There Everyone," Facebook Workplace post, January 7, 2021.

239   "It's worth stepping back": Schroepfer, "Hang In There Everyone."

239   It was a "rallying point": "Stop the Steal and Patriot Party."

240   "We were able to nip": "Stop the Steal and Patriot Party."

241   On January 7, as most: Andrew Bosworth, "Demand Side Problems," Boz .com, January 7, 2021.

241   Cracking down too hard: Andrew Bosworth, "Demand Side Problems," Facebook Workplace post, August 2020.

242   One of those files: "Stop the Steal and Patriot Party."

244   "Information Corridors could": Author name redacted, "Information Corridors: A Brief Introduction," Facebook memo, March 2021.

244   Other documents that Haugen: "Stop the Steal and Patriot Party."

246   "I think that if someone": Mark Zuckerberg interview by Mike Allen, "Mark Zuckerberg Interview," Axios and HBO, 2020.

246   A researcher randomly: Author name redacted, "Vaccine Hesitancy Is Twice as Prevalent in English Vaccine Comments Compared to English Vaccine Posts," Facebook memo, March 2021.

246   Additional research found: Author name redacted, "Directional Guidance on Covid Workstreams: 1-Week Sprint Learnings," Facebook report, April 2021.

246   "We found, like many": "Directional Guidance on Covid Workstreams."

248 Though the argument was favorable: Author name redacted, "Overview of Polarization Research," Facebook report, June 2021.

248 Rather than vindicating: "Overview of Polarization Research."

249 Titled "What We Know": Pratiti Raychoudhury and Chris Cox, "What We Know About Polarization," Facebook Workplace memo, April 2021.

250 "The more misinformation": Author name redacted, "An Echo Chamber of Trust: Information on Instagram," Facebook Workplace memo, September 24, 2020.

CHAPTER 16

259 Among them was a 2019: Author names redacted, "Hard Life Moments— Mental Health Deep Dive," Instagram presentation, November 2019.

CHAPTER 17

263 Internal research done in 2019: Author names redacted, "Problematic Use of Facebook: User Journey, Personas & Opportunity Mapping," Facebook report, March 2020; author names redacted, "Problematic Facebook Use: When People Feel Like Facebook Negatively Affects Their Life," Facebook report, July 2018.

264 In 2020, Instagram's: Instagram Well-Being team, "Social Comparison: Topics, Celebrities, Like Counts, Selfies," Instagram survey, January 2022.

264 As a fellow employee: "Social Comparison."

265 An executive at the Jed Foundation: Allyson Chiu, "Will Hiding Likes on Instagram and Facebook Improve Users' Mental Health? We Asked Experts," *Washington Post,* May 28, 2021.

265 The company had done research: Author name redacted, "Project Daisy Launch Discussion Draft Review," Facebook review, February 2020.

266 Hegeman, the head of News Feed: "Project Daisy Launch Discussion Draft Review."

266 The project was shelved: Ryan Mac and Craig Silverman, "Facebook Is Building an Instagram for Kids Under the Age of 13," *BuzzFeed News,* March 18, 2021.

268 Haugen had shared with me: Author name redacted, "Proactive Risk Detection: Eating Disorders," Facebook report, March 2021.

270 "Our product instincts told us": Julie Zhuo, "The Power of Product Thinking," Future, June 15, 2021, https://future.com.

270 During any given week: Author name redacted, "Providing Negative Feedback Should Be Easy (And Why This Would Be Game Changing for Integrity)," Facebook report, September 2019.

CHAPTER 18

284 Only after we published: "To Treat Users Fairly, Facebook Must Commit to Transparency," Oversight Board, September 2021, www.oversightboard.com/news.

287 In the weeks before we pulled the trigger: Senator Richard Blumenthal (@SenBlumenthal), "@MarshaBlackburn & I are calling on Mark Zuckerberg to release Facebook's internal research," Twitter, August 4, 2021, https://twitter.com/SenBlumenthal.

288 "We stand by this research": Karina Newton, "Using Research to Improve Your Experience," Instagram, September 14, 2021, https://about.instagram.com/blog.

289 "I'm biased, obviously, where I sit": Adam Mosseri interview by Peter Kafka, "Instagram Boss Adam Mosseri on Teenagers, Tik-Tok and Paying Creators," *Recode* podcast, *Vox*, September 2021.

289 "Instagram Chief Takes Heat": Kenneth Garger, "Instagram Chief Takes Heat for Bizarre Analogy Defending Social Media," *New York Post*, September 16, 2021.

289 "Instagram Boss Says Social Media": Jack Morse, "Instagram Boss Says Social Media Is Like Cars: People Are Going to Die," *Mashable*, September 16, 2021.

291 Titled "What the Wall": Nick Clegg, "What the Wall Street Journal Got Wrong," Meta, September 18, 2021, https://about.fb.com/news/2021.

291 The idea of an Instagram for children: Ryan Mac and Craig Silverman, "Facebook Is Building an Instagram for Kids Under the Age of 13," *BuzzFeed News*, March 18, 2021.

295 "I am begging conservatives": Dan Bongino, *The Dan Bongino Show*, October 6, 2021.

295 The *Daily Wire* pronounced: Luke Rosiak, "Facebook Whistleblower Is Leftist Activist Repped by Lawyer for 'Whistleblower' Behind Trump Impeachment," *Daily Wire*, October 5, 2021.

298 "Facebook shares rise": Salvador Rodriguez, "Facebook Shares Rise as Investors Focus on Earnings Beat and Look Past Whistleblower Document Dump," CNBC, October 25, 2021.

299 "The metaverse is the next frontier": Mark Zuckerberg, "Founder's Letter, 2021," Meta, October 28, 2021, https://about.fb.com/news/2021.

300 Even after the rebrand: Facebook internal metrics dashboards, 2021.

300 Contrary to "our prior": Author name redacted, "Meta Brand Sentiment Review," Meta review, August 2021.

CHAPTER 19

301 "I'm sure many of you have found": Mark Zuckerberg, "I wanted to share a note I wrote to everyone at our company," Facebook, October 5, 2021, www.facebook.com/zuck.

303 The nature of those: Pratiti Raychoudhury et al., "On Narrative Excellence," Facebook memo, November 1, 2021.

303 One memo required researchers: Raychoudhury et al., "On Narrative Excellence."

305 "Fundamentally, the trust & safety": Samidh Chakrabarti (@samidh), "Fundamentally, the trust & safety solution space is far broader when platforms prioritize reduction of user harm," Twitter, September 13, 2021, https://twitter.com/samidh.

305 "especially difficult for me": Samidh Chakrabarti (@samidh), "Today's WSJ reporting was especially difficult for me to read," Twitter, September 16, 2021, https://twitter.com/samidh.

305 "In my view, it's actually": Samidh Chakrabarti (@samidh), "In my view, it's actually not uncommon for FB researchers working on societal issues to feel ignored," Twitter, September 18, 2021, https://twitter.com/samidh.

306 In a story about its launch: Issie Lapowsky, "They Left Facebook's Integrity Team. Now They Want the World to Know How It Works," *Protocol*, October 26, 2021, www.protocol.com.

307 Further reporting led to: Jeff Horwitz and Katherine Blunt, "Instagram Connects Vast Pedophile Network," *Wall Street Journal*, June 7, 2023.

309 "I don't normally point out": Mark Zuckerberg, "Look, it's one thing for the media to say false things about my work," Facebook, September 21, 2021, www.facebook.com/zuck.

309 "When our work is being": Keach Hagey, Georgia Wells, Emily Glazer, Deepa Seetharaman, and Jeff Horwitz, "Facebook's Pushback: Stem the Leaks, Spin the Politics, Don't Say Sorry," *Wall Street Journal*, December 29, 2021.

311 Fully 70 percent: Author name redacted, "The Quality of Widely Viewed Content Has Improved: Evaluating the Q3 WVCR," Facebook review, October 2022.

311 "the usual growth vs. integrity tension": "Social Media & the January 6th Attack on the U.S. Capitol: Summary of Investigative Findings," U.S.

House Select Committee to Investigate the January 6th Attack on the United States Capitol.

312 By the end of 2021: Author name redacted, "The Long-Term Effects of FB App Integrity Interventions," Facebook report, 2021.

313 In one memo that leaked: Vishal Shah, "Announcing the Horizon Quality Lockdown," Meta memo, September 2021.

314 A later Workplace: Author name redacted, "JOIN THE OPEN ENROLL-MENT FAIR...IN THE METAVERSE!," Facebook Workplace post, October 2022.